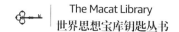

The Macat Library
世界思想宝库钥匙丛书

解析巴鲁赫·斯宾诺莎
《伦理学》

AN ANALYSIS OF
BARUCH SPINOZA'S
ETHICS

Gary Slater　Andreas Vrahimis ◎ 著

杨阳 ◎ 译

上海外语教育出版社
外教社 SHANGHAI FOREIGN LANGUAGE EDUCATION PRESS

MACAT

目　录

CONTENTS

引言

要点

- 巴鲁赫·斯宾诺莎（1632—1677），后改名为本尼迪克特·德·斯宾诺莎，荷兰哲学家。因其著作《伦理学》过于激进，斯宾诺莎所在的犹太社区曾将他革出教门*。但如今《伦理学》已被视为现代哲学的要籍。

- 《伦理学》认为，万物最终都是一个单一实体的表达，斯宾诺莎将其称为神，等同于自然。他反对神人同形同性论*这一普遍观点（即人格化的神，神有人性，故而爱人或惩罚人）。

- 人类如何在一个没有目的且自己并不处于首要地位的世界幸福生活？《伦理学》使得这种思索成为可能。

巴鲁赫·斯宾诺莎其人

巴鲁赫·斯宾诺莎于 1632 年出生于阿姆斯特丹。他的家人是马拉诺人*——因遭受迫害而逃离葡萄牙的犹太人。他在阿姆斯特丹的马拉诺社区接受了教育。1656 年，显然是因为他饱受争议的神学观点，该社区将其革出教门。斯宾诺莎从此过着相对封闭的生活，在莱茵斯堡的小镇上以磨镜片为生，后又迁居福尔堡、海牙等地。

被驱逐之后，斯宾诺莎将犹太名（巴鲁赫）改为拉丁文形式的本尼迪克特，并以此为他的大部分哲学著作署名。值得一提的例外是 1670 年匿名出版的《神学与政治学专题研究》，这部作品是斯宾诺莎生前出版的著作中最受争议的一部。1674 年荷兰归正会宗教议会*（即总议会）查禁了这部作品，在那时作者已经确定是斯宾诺莎了。

《伦理学》于斯宾诺莎死后出版，甫一面世便遭到几乎一致申讨。很长一段时间内，哲学家们都试图与斯宾诺莎的激进观点保持距离。直到斯宾诺莎逝世大约一个世纪之后，他的著作才在哲学争论中得到正视。

斯宾诺莎的恶名某种程度上来自他对某些原则的坚持，而这些原则却不被当时所容。但一些他生前所争取的原则，包括他对宗教宽容和对公民基本自由的捍卫，已在当今西方世界得到了普遍认同。

《伦理学》的主要内容

文艺复兴*晚期，随着现代科学的崛起，人们发现地球并不是宇宙的中心。这一颠覆性的发现迫使人类重新定位自己在宇宙中的位置。人类生存在一个受自然决定论*制约的世界，换言之，恰如严格的科学法则支配着自然世界，人类世界也由一定的法则支配着，并且这些法则不以人类的自由意志为转移。斯宾诺莎的《伦理学》就是这个世界的生存指南。如何在一个由决定论支配的世界幸福地生活？斯宾诺莎为了得到这个问题的答案，历经了不同的阶段。《伦理学》的总体论证基于由古希腊数学家欧几里得*首创的几何学方法*：斯宾诺莎首先将自己的一系列术语的定义与一系列定理*（不证自明的命题）相结合，由此他试图演示一系列命题的证明过程。这些证明涉及哲学和其他学科的诸多话题，从形而上学*（对存在本质的研究）和神学到认识论*（对知识的本质和局限的探究）、心理学以及行动理论（斯宾诺莎对行动之意义的分析），最后解释了何为人类的自由。

在斯宾诺莎的形而上学中，他发展出了一种泛神*一元论*，并认为神即自然。（一元论认为所有存在都是唯一本原的表现形式；泛神论将这一本原等同于神。）他认为整个现实都是神的表现形式，神才是所有现实之下的本原，斯宾诺莎用"实体"来指称这一本原。斯宾诺莎所言之神（或自然）就是这种单一实体，它不以人的形象出现。神不能被人格化（即神不具人性），于是与各家宗教的观点不同，神并不具备爱人的能力，也不会因为人的罪孽而降下惩罚。

斯宾诺莎从他的一元论得出自由意志不存在的推论。他认为人只有承认世界是在决定论（即每一个事件都是一个使之必要和必然的原因所致）的制约下运转的，人类才能得到自由。基于这种前提，《伦理学》对人的情感的分析很有新意，它认为人的情感源自喜悦、悲伤和欲望。反过来，斯宾诺莎认为这些情感也是努力坚持存在（即一种生存下去和获得成功的动力）的表现。他认为幸福的生活来自人对神的理智的爱，纵然神是不会反过来爱人的。这种对神的理智的爱需要对决定论式的因果关系 * 的认识与理解。

对于其同时代的人来说，斯宾诺莎的思想过于激进。否认自由意志，反对神人同性，这些观念都不是 17 世纪的思想家所能接受的。18 世纪晚期德国哲学界关于泛神论的一场辩论之后，斯宾诺莎的思想才初具影响，才有后来格奥尔格·威廉·弗里德里希·黑格尔 * 宣称："你要么是个斯宾诺莎的信徒，要么就根本不是一个哲学家。"[1]

《伦理学》的学术价值

斯宾诺莎将《伦理学》写得相当费解，尽管如此，想要跟上形而上学和认识论的现代讨论进展的哲学专业学生必读此书。比如，其中对于身心关系的讨论就极具原创性和重要性。斯宾诺莎将身体和心灵视为同一本原的两种不同表现形式。按照他的观点，任何事物都有身体和心灵的双重表达。

对于我们现在普遍称为"情感"的心理机制，斯宾诺莎也有新颖的见解，他称之为"情状"。《伦理学》定义并且深入讨论了爱、恨、希望、恐惧及其他情状。斯宾诺莎认为它们全部源自三种基本情状——喜悦、悲伤和欲望。这些情状是身体状态在心灵中的反映，它们也会进一步占据心灵，使人不能理性思考。斯宾诺莎认为了解这些情感状态背后的运作机制就能让我们控制它们，进而也为幸福生活打下了基础。这样的观点反映出古代斯多葛派 * 思想对于斯宾诺莎的影响。斯多葛派学者也认为调整好情感才能过上美好的生活。

这就是斯宾诺莎著作中不同层次的分析的主要内容。《伦理学》从对形而上学的抽象讨论开始，最终落在对伦理学的具体讨论上，中间涉及心灵哲学、认识论、心理学、行动理论和对自由的探讨。斯宾诺莎思想的方方面面影响了不同领域的人们。在斯宾诺莎的影响下，许多新思想或可显现。

---

1. 默罗阿德·韦斯特法尔："斯宾诺莎与德里达之间的黑格尔"，大卫·杜奎特编，《黑格尔的哲学史：新阐释》，奥尔巴尼：纽约州立出版社，2003 年，第 144 页。

第一部分：学术渊源

1 作者生平与历史背景

要 点 🔑

- 斯宾诺莎的这部著作是现代哲学的关键文本。它要处理的问题是：在一个人类行为受决定论制约而非自由意志指导的世界里如何过上美好的生活。

- 因与阿姆斯特丹的犹太团体的既有观念相左，斯宾诺莎遭到驱逐。他迁居荷兰的其他地区，写下了许多伟大的作品。

- 斯宾诺莎支持 17 世纪荷兰的自由主义。自由主义政府在 1672 年被保皇势力推翻后，荷兰的自由主义也遽然终止。

为何要读这部著作？

本尼迪克特·德·斯宾诺莎的《伦理学》是哲学史上最重要的著作之一。这本书是用拉丁文写成的，创作于 1661/1662 年与 1675 年之间，在作者逝世后于 1677 年出版。这本书涉及对人类行为的是非对错的讨论，书名即来源于此。但这个书名会误导读者，因为《伦理学》也是一部关于神和世界的书，它检视了心灵的结构和心灵与身体的关系，详述了各类情感的心理机制，并试图去理解人的行为。《伦理学》否认自由意志，探索决定论（一种哲学观点，它认为人的所有行为都是由自由意志以外的原因决定的）的影响，时至今日这仍是一种激进的观点。这本书认为利他主义 *（为他人的幸福着想）是追求自身利益的结果，而不是与自身利益相悖。它深入分析人的情感，得出美好生活来源于愉悦和行动力这一结论。阅读这本书会很吃力，但它无疑是现代哲学的关键文本。

《伦理学》提出的问题包括：世界是什么？我们如何认识世界？我们的情感是怎么与之相关的？在这样的世界中我们该如何行动？不管是在斯宾诺莎的时代还是在今天，这些问题都同样重要。《伦理学》解决这些问题的方式系统且新颖，这就是它的长久价值所在。甚至这本书的几何学风格也不是随意为之。这种风格体现了斯宾诺莎对自然规律的信仰，它不偏不倚，面面俱到，对万物都具有同等效力。

> "根据天使的判决、圣人的指令，我们革除、驱逐、诅咒并谴责巴鲁赫·德·斯宾诺莎……主会把他从人间除名……但是今日笃信主的各位是活着的。我们命令：任何人不得与其说话或通信、不得可怜他、不得与之同住、不得接近他四尺之内，凡他所著所写均不可阅读。"
>
> ——"阿姆斯特丹塔尔穆德塔拉教会对斯宾诺莎的驱逐令"，
> 引自斯蒂芬·纳德勒《斯宾诺莎传》

作者生平

斯宾诺莎 1632 年生于荷兰阿姆斯特丹，希伯来名是巴鲁赫。他成长于阿姆斯特丹的马拉诺人社区，打理过一段时间的家族进口生意，与此同时他也展现了自己做学者的潜力。人们对斯宾诺莎的早年生活知之甚少；主要由于他与众不同的神学观点，那段岁月颇不平静。可能是因为他如此离经叛道，他在 1656 年遭到了持刀袭击。[1] 同年 6 月 27 日，他所在的社区发公文称其为"当灭之物"*（驱逐社区成员时的希伯来语术语），将其革出教门，接着又指控他宣传"可怖的异端邪说"和有"残暴行为"。[2] 驱逐他的具体原因并不清楚，可能是由于他对预言、永生和神的观点与正统相背离。他的家人、同事、同窗都被禁止与他联系。1661年，斯宾诺莎离开阿姆斯特丹，前往荷兰莱顿附近的莱茵斯堡村，在那里他开始写作《伦理学》。同时，他将自己的名字拉丁化，改为本尼迪克特。斯宾诺莎先住在莱茵斯堡，后迁居福尔堡、海牙等地，靠给光学仪器打磨镜片来支持自己的哲学工作。

遭驱逐期间，斯宾诺莎与一群朋友和支持者通信往来，他们帮斯宾诺莎树立名望，并传阅《伦理学》的初稿（至少在荷兰和英格兰两地他们成功了）。这些朋友和支持者中的许多成员属于门诺派*，这是一个非主流的新教教派，他们引起了政府当局和教会的怀疑。在17世纪的欧洲，特别是在英格兰和荷兰，当权者对某些知识分子团体的敌意越来越强（但尚未泛化到文化层面）。斯宾诺莎的朋友鼓励他出版《伦理学》，但他有所犹豫。一方面是因为被革出教门的创痛，一方面是因为他的另一部著作所遭受的敌意，这本书就是1670年匿名出版的《神学与政治学专题研究》，它和《伦理学》有许多相似的观点。

1677年2月，斯宾诺莎因肺结核逝世。死因疑似吸入玻璃粉末，这是他从事打磨镜片工作的后果。同年，《伦理学》在他死后出版。

创作背景

斯宾诺莎创作和出版作品的时代是荷兰历史上的一段动荡岁月，当时的荷兰共和国由扬·德·维特*执掌，重视政治上的自由主义和宗教宽容。事实上，斯宾诺莎在17世纪70年代德·维特统治的全盛时期领取过他特批的一小笔津贴。[3] 荷兰的总体气氛有利于传播自由主义观念，也有利于出版斯宾诺莎颇有争议的作品。

《神学与政治学专题研究》(1670 年首次出版) 中即提出了许多引起争议的观点，这本书可以被解读成对德·维特政治计划的辩护。但值得一提的是，即便在这样宽松的氛围中，斯宾诺莎仍惴惴不安，不愿用真名发表有争议的观点。

很快，斯宾诺莎的谨慎就被证明是正确的，因为他的作品的命运还是和德·维特的命运绑定了。1672 年（对荷兰来说，这一年是"灾难之年"），法荷战争之后，德·维特及其追随者被保皇党推翻并处以私刑。斯宾诺莎无法继续匿名写作，这样的气氛下他的作品很快就遭到攻击，被称为是"犹太教的叛徒和魔鬼在地狱里炮制的，在扬·德·维特的授意下发表的"，[4]因此荷兰归正会在 1674 年声讨和查禁了他的作品。[5]《伦理学》出版于斯宾诺莎逝世后，在接下来的一个世纪里这本书也遭受了相似的攻击。在这一时期，哲学家们普遍以追随斯宾诺莎为耻。直到 18 世纪晚期，经过好一阵犹疑之后，人们才将斯宾诺莎的作品纳入严肃的哲学讨论范围内。

---

1. 初载于比埃尔·培尔所著的斯宾诺莎传记；参见 H. M. 拉文和 L. E. 古德曼：《斯宾诺莎哲学中的犹太主题》，纽约：纽约州立大学出版社，2012 年，第 269 页。

2. 吉纳维芙·劳埃德：《劳特利奇哲学导论·斯宾诺莎与〈伦理学〉》，伦敦：劳特利奇出版社，1996 年，第 1 页。

3. 参见罗杰·史克鲁顿：《斯宾诺莎》，牛津：牛津大学出版社，1986 年，第 11 页。

4. 史克鲁顿：《斯宾诺莎》，第 11 页。

5. 史克鲁顿：《斯宾诺莎》，第 11 页。

2 学术背景

要点 ☗━━

- 斯宾诺莎的思考反映了他那个时代的大事。欧洲的宗教战争让他重视宗教宽容的理念，科学革命的勃兴让他质疑宗教对世界的解读。

- 斯宾诺莎的思想主要受到了勒内·笛卡尔*的影响。笛卡尔提出心灵和身体是独立存在的。但是在后世的哲学家看来，他关于二者如何互动的理论存在不足之处。

- 斯宾诺莎否认亚里士多德*的"目的因"*。他认为心灵和身体是同一实体的两个方面（而不是截然不同的两个实体），这与笛卡尔是不同的。

著作语境

本尼迪克特·德·斯宾诺莎的《伦理学》创作于欧洲思想史上的一段变革时期。在这之前，宗教改革*引发的战争席卷了整个欧洲大陆，宗教宽容的观念也在斯宾诺莎的居住地荷兰逐渐形成。现代科学的各种发现也在逐步取代旧的中世纪经院哲学*体系，后者与罗马天主教会息息相关。许多原本确定的概念（例如地心说，认为地球是宇宙的中心）瓦解了，经院哲学也随之终结。

当时的政治发展和现代科学的质疑精神都影响了斯宾诺莎。他深受宗教迫害之苦，认为宗教宽容有无上的价值，更愿意用一生去为之辩护。斯宾诺莎的宗教观浸透着现代科学带来的新的质疑精神，也源于当时人们发现自然世界可以通过自然的决定论法则来解释。决定论法则规定什么样的原因必然造成什么样的后果。斯宾诺莎思想的基本特征是信奉决定论，同时还研究人如何在一个由决定论支配的世界里过上美好的生活。

> "最终我发现了它——思想，只有它和我是无法分离的。我是，我存在——那是当然……于是，在严格的意义上我只是会思考的一个东西罢了；也就是说，我是心灵，或是智力，或是智识，或是理性——这些词语的意义我到如今还是不甚明了。但尽管如此，我仍是个真的、确实存在的东西。但是哪种东西呢？如前所述——一个会思考的东西。"
>
> ——勒内·笛卡尔：《第一哲学沉思录》

学科概览

斯宾诺莎及其同辈哲学家写作的时代适逢哲学界开始偏离旧的亚里士多德哲学传统。造成这种偏离的就是现代科学的兴起。这场运动的领军人物是勒内·笛卡尔，他是西方现代哲学的奠基人。笛卡尔认为心灵与物质不能混为一谈。他将物质世界称作"广延物"*，意思是"有广延性的东西"，因为他认为空间上的延伸这一特性才是物质的主要特征。广延物存在于心灵领域（思想物*，能思考的东西）之外。对笛卡尔来说，心灵和身体是截然不同的两个实体，彼此独立存在。顺着这个观点笛卡尔进而声称：按照他的物理学，一切实物均可以在力学规律的基础上加以解释。他认为，物质实体就像是一架机器里的齿轮，一个零件的运动不可避免地会带动另一个。物质世界受决定论的法则支配，这种法则可以通过物理学去发现。

笛卡尔的身心二元论暗示：虽然身体可以用物理学法则来解释，心灵却不由这种决定论法则支配；所以和身体相反，心灵可以拥有自由意志。但是笛卡尔的说法一开始就存在着一个严重的问题：既然心灵和身体是相互独立的实体，那心灵怎么能与身体互动呢？我想移动我身体的一部分（也就是说，仅仅因为我的心灵能够自由选择这么做），就能够移动，这是怎么做到的？

笛卡尔试图这样回答：身体的一部分位于大脑中，它是可以与心灵互动的。但是这样的答案并不让人满意，因为它利用的是身心二元论中的一个例外情况。针对这个所谓的身心问题，笛卡尔之后包括斯宾诺莎在内的现代哲学家都试图给出其他的解答。

学术渊源

斯宾诺莎的童年教育深深地影响了他。他受到的教育包括学习犹太教传统和了解一些哲学家，比如 12 世纪犹太教圣经学者迈蒙尼德＊和古希腊哲学家亚里士多德。除了在他成长的社区接受教育之外，斯宾诺莎还跟随弗朗西斯科斯·范·登·恩登＊学习拉丁文。恩登是前耶稣会士，也是一个政治激进分子，当时很多人文主义者和自由思想家聚集在他家中，斯宾诺莎的很多政治观点来源于这里。

就《伦理学》的内容来说，最好是把现代和前现代的影响分开来看。斯宾诺莎关于法律的哲学基础的观点有迈蒙尼德的影响。同样也是在迈蒙尼德的影响下，斯宾诺莎形成了这样的观点：人类不是造物的中心，神既是理解的来源，又是被理解的总体。另一个《伦理学》的前现代影响是古典斯多葛派思想。斯多葛派认为，人们应该平静地接受他们无法掌控的东西，《伦理学》的最后三个部分的思想就来源于此。第三,《伦理学》在处理因果关系和实体时所使用的术语来自亚里士多德哲学,尽管二者所得出的结论是截然不同的。亚里士多德区分了两种不同类型的因果关系:动力因(即我们今天所认为的显而易见的因果关系,例如,一艘船存在的动力因是造这艘船的人)和目的因(一个东西存在的目的,例如,一艘船存在的目的因是在水上航行)。斯宾诺莎认为只有动力因才是真正的原因,并且否认目的因的存在。由此,他展开了对目的论 * 的激烈批判,因为目的论就是研究事物目的的学说。亚里士多德认为世界是由许多不同的实体组成的,斯宾诺莎却认为一切事物只不过是一个单一实体的表现形式而已,这个实体就是神,也就是自然。

这本书所受的现代影响中最突出的来自笛卡尔。斯宾诺莎是研究笛卡尔的权威专家,他在《伦理学》中所使用的来自笛卡尔的术语要远远多于其他任何思想家。然而,就像使用亚里士多德的术语一样,斯宾诺莎使用这些术语所得出的结论与笛卡尔所得出的结论也迥乎不同。笛卡尔所说的神是一种独特意义上的实体,以一种独一无二的方式独立于任何其他存在;他始终认为世界是一片由理性法则所支配的领域。然而斯宾诺莎却把神这个概念转换为一个独特的实体,所有其他的事物都只是这个实体的样式或表现形式而已。

3 主导命题

要点 🗝

- 随着现代科学的兴起，与"人类是宇宙中心"有关的一系列观念开始引起人们的质疑，比如人与其他的存在不一样，人具有自由意志。

- 当时，现代科学家和哲学家们，比如勒内·笛卡尔，团结一致对抗欧洲大学里相对保守的经院哲学思想，斯宾诺莎就是在这样的时代开始了自己的写作。

- 对于斯宾诺莎和当时其他的理性主义*哲学家来说，一个尚未解决的关键问题是如何解释心灵和身体之间的互动，这两者被认为是两个独立的实体。

核心问题

如何在一个没有目的、没有自由意志、人类并不处于首要位置的世界行动？这是本尼迪克特·德·斯宾诺莎的《伦理学》所要处理的核心问题。斯宾诺莎给出的解答基于这部作品最具独创性的观念——"欲求"*，即拉丁文的"努力"，意思是为存在而付出的努力（即努力生存并发展）。斯宾诺莎提出，一个人若想过上美好的生活，就要听从欲求行事，遵循自身利益所向。

斯宾诺莎的伦理学观念基于人人利己的判断，就是说每个人都在为了自己努力生存。对斯宾诺莎来说，这样的利己是种美德。值得注意的是，对他来说，人应该运用智识去生存，而不是靠生理本能去求生。因此，欲求与了解或理解事物的努力紧密相关。虽然这样貌似是放纵纯粹的自私，但斯宾诺莎的本意是借此来解释为什么利他主义实际上是追求自我利益的产物。在斯宾诺莎看来，一个追求美好生活的利己主义者也一定是利他的。换句话说，利己主义者关心他人的幸福大有必要。

斯宾诺莎的"欲求"这一概念引出的另一个结论是，自由意志是一种假象。有人认为有意识的行为就是心灵控制了身体的一种表现，而斯宾诺莎认为至少在这一层面上自由意志是一种假象。就像他说的，"心灵中不存在绝对的或者自由的意志，心灵是被某种原因推动而决定了去做这个或做那个，这个原因又被另一个原因推动着，如此类推，没有尽头。"[1]

> "善恶就其本身来说并不说明事物的好坏，而只是思考的样式而已，即我们将事物相互比较而产生的看法。同样一个事物，可以同时是好的、坏的和无关的。比如人在心情低落时会觉得音乐悦耳，但是在哀悼亲人时就不会欢迎音乐了；而对一个耳聋的人来说，音乐是没有意义的，无所谓好坏。"
>
> —— 本尼迪克特·德·斯宾诺莎：《伦理学》

参与者

斯宾诺莎写作《伦理学》时，知识界的主要冲突来自经院哲学家和现代思想家对欧洲大学命运的争夺。自中世纪起，大学中的思想生活就由经院哲学掌控，经院哲学家喜欢针对亚里士多德和其他权威作者著作中的逻辑和形而上学问题（研究存在本质的学说）进行冗长辩论。

到了 17 世纪，人们对探索自然越来越有兴趣，并对欧洲大陆充满暴力的宗教冲突感到恐惧，这都促使他们去探索建立知识体系的新方法。这些探索注重观察与试验，以及力求普适（超越文化差异）的传播风格。斯宾诺莎同他的前辈勒内·笛卡尔一道坚定地与现代哲学家为伍，《伦理学》就处于这次思想转型的最前沿。到斯宾诺莎开始写作时，现代笛卡尔式的理性主义已经开始作为有别于经院哲学的另一种哲学研究方法被建立起来了。与经验主义*（认为一切知识只能靠观察获得）不同，理性主义认为有些知识单凭推理即可得出。除了发起这项运动的笛卡尔之外，理性主义哲学的主要代表人物还有尼古拉斯·马勒伯朗士*以及较晚期的戈特弗里德·威廉·冯·莱布尼茨*。

当时的论战

在斯宾诺莎生活和写作的时代，现代科学的兴起已经开始引导人们质疑亚里士多德的经院哲学。"目的因"就是亚里士多德的一个受到质疑的中心概念。根据亚里士多德的构想，一个人要想完全理解一个现象，需要理解这个现象的目的或者其"目的因"（希腊语作 telos），即"它是做什么用的？"这一问题的答案。因此，一位信奉亚里士多德的自然哲学家会认为对目的的研究（即目的论）是所有科学的至要。现代科学不会认可这种目的论式的解释，例如笛卡尔认为自然的决定论（即他所说的"机械论"）法则可以完全解释所有物质事物，根本不需要考虑它们的目的。但笛卡尔也认为，除了物质事物，还存在着精神事物，后者是不遵从机械论定律的。

斯宾诺莎形而上学的基础是他对"实体"这一概念的讨论，他在这一方面受到笛卡尔的直接影响。笛卡尔彻底修正了之前亚里士多德对实体的定义，也把这一概念引入现代哲学的讨论中。对笛卡尔来说，一个实体并不依存于其他事物的存在而存在，他把实体分为两种：无限的和有限的。在笛卡尔看来，只有一种实体是无限的，它的名字叫做神。神是自因的，他的存在并不依存于其他实体。相反，一个有限实体的存在只取决于无限实体的存在，也就是创造它的神。一个有限实体一旦被神创造出来，它的存在就独立于任何其他的实体了。笛卡尔认为有两种有限实体：心灵和身体。他认为就算身体不存在，心灵也能够存在，反之亦然。

于是这在当时引发了一场理性主义哲学家之间的辩论：心灵和身体这两个截然不同的实体是怎么互为因果的呢？比如，我想举起胳膊怎么就成了我举起胳膊的原因呢？既然笛卡尔认为心灵和身体是分开的，这个问题就很难回答了。他试图这样回答：大脑中的某一部分会通过某种不为人知的方式让心灵指挥身体。但这又与他所坚持的身心二元论相矛盾了——因为这个回答似乎暗示了既是身体又是精神的事物是存在的。有些信奉笛卡尔主义的哲学家试图这样解决这个问题：他们否认心灵和身体之间有一种互为因果的关系，这个观点也得到了斯宾诺莎的支持。

---

1. 本尼迪克特·德·斯宾诺莎：《伦理学》，选自《斯宾诺莎读本：〈伦理学〉及其他著作》，埃德温·柯利译，普林斯顿：普林斯顿大学出版社，1996 年，第 146 页。

4 作者贡献

要点 🔑

- 《伦理学》教导我们，要想在一个没有自由意志的世界过上幸福生活，人类需要理解并调节自己的心理状态。

- 斯宾诺莎认为欧几里得的几何学才能最好地表达出他的哲学的精妙之处。他把神定义为"绝对的无限"。

- 斯宾诺莎的《伦理学》的基础是勒内·笛卡尔引入的理性主义框架，但是笛卡尔的模式允许人们有自由意志，而斯宾诺莎的决定论却不允许。

作者目标

在一个对我们的担忧毫不在乎的由决定论支配的世界里为人类的幸福指出一条路来，这就是本尼迪克特·德·斯宾诺莎写作《伦理学》的根本目标。达到这个目标需要清晰地理解物质或精神事物的原因，然后才能对这个世界有尽可能清晰的把握。斯宾诺莎心理学的基础是关于形而上学和认识论（哲学的两个分支，分别研究存在的概念与知识的本质和局限）的论述。随后，斯宾诺莎心理学也引出了他的行为理论，以及他对如何过上美好生活的总体论述。提出一套关于美好生活的论述就成为他的终极目标，为实现这一终极目标，需要处理上述几个领域的严肃问题。《伦理学》从处理最抽象的问题出发，接着再系统性地处理细节问题。

于是《伦理学》开头先对实体的本质提出了基本的质疑。斯宾诺莎想要先厘清前辈哲学家针对这个话题的论述中的几个问题，特别是勒内·笛卡尔的论述，根据他的实体概念，笛卡尔声称心灵和身体是两个截然不同的实体。这导致了一个难题：心灵和身体之间如何互动并对彼此产生影响？斯宾诺莎试图这样解决这个问题：他认为心灵和身体之间没有因果联系（其中一方并不能引起另一方的行动和改变）。《伦理学》对笛卡尔论述中的问题进行了有趣的重塑，这造成的结果之一就是斯宾诺莎接受了决定论。

斯宾诺莎把他对实体的重新定义和对笛卡尔问题的解释延伸到了认识论的领域。他的目标是提供一种关于知识的理性主义论述（也就是说，同心灵自身的知识相比，通过感觉所获得的知识可能是不完整的）。这种论述巩固了斯宾诺莎的心理学，后者试图说明我们的心灵对我们的情感原因的了解是如何成为在一个决定论的世界里幸福生活的基础的。

> "但是我想我已经充分说明了无限多的事物在无限多的方式下都自神的无上威力或无限本性中必然流出；这就是说一切事物从永恒到永恒都以同等的必然性自神而出，正如三内角之和等于二直角是从三角形的必然性而出那样。"
>
> —— 本尼迪克特·德·斯宾诺莎:《伦理学》

研究方法

斯宾诺莎非常重视规范的哲学表达，他觉得用数学来完成这种表达是最好的。他认为欧几里得的几何学是一种表达哲学思想的理想方法，要"把人的行动、欲望都当成线、面、体的问题"。[1] 他写书就好像在写几何学文本，先从一组术语的定义开始。例如，他把神定义为"绝对无限的存在，即由无限种特性所构成的实体，每一种特性都表达出一种永恒的和无限的本质"。[2] 这个定义也是基于之前"实体"和"属性"的定义而得出的。定义了基本术语之后，斯宾诺莎接着给出的是定理 *（不证自明的论断），比如"不管是什么，要么在自身之中要么在他物之中。"[3]

和在几何学论文里一样，斯宾诺莎的定理也被设想为不证自明的真理，其本身是无法被证明的。这些定理和定义是接下来所有证明要遵循的基础。剩下的所有内容是由证明、论证和评注 *（对个别论证和证明的解释性评论）构成的。这些证明要么纯粹通过参考定义和定理得到论证，要么就参考前文中已经成立的证明。因此，文本内容的推进是建立在之前的材料上的。

时代贡献

虽然斯宾诺莎认为这些定义和定理是普遍而永恒的，但其中经常会有一些观点在他看来是特别适合当时的哲学家的。比如，他对"实体""特性""样式"这些术语的定义依赖于笛卡尔对这些术语的使用。[4] 斯宾诺莎使用了笛卡尔哲学中的词汇，但是修正了一些基本概念，目的是为了解决一些笛卡尔提出的基本问题。斯宾诺莎使用笛卡尔关于"实体"的定义，是为了论证笛卡尔自己使用这个术语有不一致或矛盾之处。斯宾诺莎首先指出，如果一个人遵循笛卡尔对于实体的定义，[5] 那么具有任何特定属性的实体不会多于一个；[6] 其次，一个实体只能由同类的另一个实体产生。[7] 这些论断是十分激进的，其中最大胆的论断是没有任何实体可以被制造出来，那么实体就是永恒的[8]、无目的的[9]。而且神必须是唯一的实体，它由无限的属性组成[10]，任何其他事物都一定是神的一种样式——或神的另一种表达形式[11]。

　　虽然斯宾诺莎从当时理性主义哲学家所熟悉的一些假设出发，但他最后得出的结论却和这些哲学家的观点相去甚远。笛卡尔试图保护自由意志这一观念不被伴随现代科学而来的决定论所消解，斯宾诺莎的"只能有一个实体"的结论却使得笛卡尔前功尽弃。笛卡尔认为决定论只适用于物质实体（即事物永远遵循物理法则），而精神实体不受物理法则的支配。斯宾诺莎指出，这一论述说明笛卡尔对于实体概念的使用前后矛盾，实体在斯宾诺莎的体系中恰恰是决定论的基础。也就是说，对斯宾诺莎而言，不存在自由意志，因为心灵也受到宇宙法则的控制。

---

1. 本尼迪克特·德·斯宾诺莎：《伦理学》，选自《斯宾诺莎读本：〈伦理学〉及其他著作》，埃德温·柯利译，普林斯顿：普林斯顿大学出版社，1996年，第 152 页。

2. 斯宾诺莎：《伦理学》，第 85 页。

3. 斯宾诺莎：《伦理学》，第 86 页。

4. 勒内·笛卡尔：《哲学原理》，V. 罗杰·米勒和 R. P. 米勒译，伦敦：里德尔出版社，1983年，第一部分，第 51—56 小节。

5. 斯宾诺莎：《伦理学》，第 85 页。

6. 斯宾诺莎：《伦理学》，第 87 页。

7. 斯宾诺莎：《伦理学》，第 87 页。

8. 斯宾诺莎：《伦理学》，第 88—90、100 页。

9. 斯宾诺莎：《伦理学》，第 97 页。

10. 斯宾诺莎：《伦理学》，第 90 页。

11. 斯宾诺莎：《伦理学》，第 94 页。

第二部分：学术思想

5 思想主脉

要点 🗝

- 斯宾诺莎不同于笛卡尔之处在于他提出的"实体一元论",他认为只存在一种"实体",并称之为"神或自然"。

- 《伦理学》开篇就是对"实体"的研究。斯宾诺莎讨论了两种定义实体的主要分类:属性和样式。

- 这本书是以欧几里得几何学为框架基础的,其理性与精确性正好与斯宾诺莎认为"宇宙不在乎人类"的观点相契合。

核心主题

本尼迪克特·德·斯宾诺莎的《伦理学》开篇先给一些基本的形而上学术语下了定义,比如"实体"、"属性"和"样式"。基于这些定义,斯宾诺莎支持"实体一元论",这一观点认为只存在一种实体,斯宾诺莎称之为"神或自然"。[1]斯宾诺莎似乎在说神和自然是同一个东西、同一个实体,这种观点叫做泛神论*。这种实体有两种基本特性,斯宾诺莎命名为心灵与身体,这遵循了笛卡尔的论述。斯宾诺莎进一步论证说,这两种特性是平行存在的,每一种事物都是这个实体的一个样式(或不同表现形式),都具有心灵和身体两种表现形式。

正文由五个相辅相成的部分组成。《伦理学》厘清了认识的本质、存在的本质和神的概念,同时认为人通过运用智力而得到自由。这些内容是《伦理学》的中心论点,书中对这些形而上学概念进行了完整论证。《伦理学》中关于认识论和行动理论的部分也同样重要。

> "我要把人的行动、欲望都当成线、面、体的问题。"
>
> —— 本尼迪克特·德·斯宾诺莎:《伦理学》

思想探究

斯宾诺莎在《伦理学》的开篇讨论了实体的本质,受笛卡尔的影响,他给出的定义是"在自身内并通过自身而被认识的东西"。[2] 换言之,斯宾诺莎认为一个"实体"不用依靠其他任何存在而存在(它的存在不以其他存在物为前提,也不被其决定)。

从这个定义出发,斯宾诺莎得出了一系列关于实体本质的结论。他否认了笛卡尔所说的"有限的实体"之间的区别,即实体(思想物和广延物)彼此之间独立存在,但都由一个"无限的实体"(笛卡尔认为的神)所创造。按照斯宾诺莎的观点,只能有一个实体,众所周知他将其等同于神或自然(神即自然)。

如读者所见,斯宾诺莎的体系着眼于实体的类别。这是一个至关重要的本体论*因素(即与关于存在的研究有关)。斯宾诺莎还使用了另外两种类别:实体的"属性"和实体的"样式"。《伦理学》一开篇就把属性这个类别定义为"由智力看来是构成实体本质的东西"。[3] 凭借这个模糊的定义去理解斯宾诺莎的意思很困难,对这一定义的诸多解释也是相互矛盾的。就实体一例来说,他的定义似乎是建立在笛卡尔关于属性的概念上。笛卡尔认为心灵的关键属性是思考,身体的关键属性是延伸——也就是说,从最基本的层面来说,身体是可以在物理空间中延伸的东西。斯宾诺莎的观点与笛卡尔相反,他认为心灵和身体不可能像笛卡尔所想的那样是分开的两个实体;他坚持认为心灵和身体是同一实体的不同属性。在斯宾诺莎看来，思考和广延是神的思考和神的广延（也就是说，他们是神这一实体的重要属性）。但令人疑惑的是，斯宾诺莎似乎暗示实体还有无穷多的其他方法来显示自身——人类碰巧只是接触到了这两种而已。[4]

除了实体和属性之外，我们在斯宾诺莎的作品里找到的第三种基本的本体论类别是样式，他的定义是"实体的分殊，亦即在他物内通过他物而被认知的东西"。[5] 和属性的定义一样，对这个定义的解读也存在着一些相同的问题。因为按照这个定义，样式是经由属性被设想出来的。任何一种思想都是能被归类于思想的普遍属性里的一种样式，任何一种身体都是属于延伸的普遍属性的一种样式。无限的样式包括支配所有思想的普遍法则和支配所有延伸的身体的普遍原理（即几何学和物理学的法则）。而有限的样式则包括所有特定的思想和身体。

斯宾诺莎的本体论体系是他解决笛卡尔关于心灵与身体关系问题的基础。在他看来，心灵与身体只不过是同一根本现实——也就是单一实体（神或自然）——的两个方面。斯宾诺莎认为，心灵和身体是这一实体的两种相互对应的表现方法：每一个事物都有其身体的一面和精神的一面。斯宾诺莎认为，即便人类心灵没有想到，每一个身体性的事物也都有一个精神性的事物与之对应：所有的精神性事物和身体性事物都是单一实体的样式（他们产生于神或自然的心灵和身体中）。于是斯宾诺莎可以通过人类与单一实体心灵和身体属性的关系来进一步解释人类的知识与行动。

语言表述

《伦理学》展开其论述的框架明显借自欧几里得几何学，因为它精确、前后一致并具有普适性。《伦理学》百分之八十的篇幅都是由定义和命题组成的复杂模型，这些定义和命题是互相指涉的——一个定义或者命题经常会指向文中的其他定义、命题。但是全书五章都包括一些文字段落，就是所谓的"评注"，语言较易理解。读者的确会在这些段落里体会到斯宾诺莎式的超然、反讽甚至幽默。这些评注可以帮助我们将一些看似迥然不同的内容统一起来。这两种表达风格——刻板的数学和微妙的反讽——之间的关系确实让《伦理学》的内容总体上更容易理解了。《伦理学》提出的论点与这些论点被提出的形式是分不开的。几何学形式不受人类影响的永恒、不变、理性的本质恰好与斯宾诺莎认为"宇宙无目的"这一观点契合。

不幸的是，对大多数读者来说，《伦理学》的数学式写作风格使它较难理解。人们认为这本书的词汇很奇怪，它的主题极为复杂，甚至令人费解。[6] 尽管阅读难度很大，《伦理学》的形式和内容合起来在公共和学术领域产生了一系列强有力的论断。在人们为了理解文化而阅读哲学史的当下，这些特点让《伦理学》变得大有裨益。

---

1. 本尼迪克特·德·斯宾诺莎：《伦理学》，选自《斯宾诺莎读本：〈伦理学〉及其他著作》，埃德温·柯利译，普林斯顿：普林斯顿大学出版社，1996 年，第 97—100 页。
2. 斯宾诺莎：《伦理学》，第 85 页。
3. 斯宾诺莎：《伦理学》，第 85 页。

4. 参见阿伯拉罕·沃尔夫："斯宾诺莎的属性概念",《斯宾诺莎研究:批评与阐释》,S.保罗·卡什亚普编,加利福尼亚州伯克利:加州大学出版社,1972年,第 26 页。

5. 斯宾诺莎:《伦理学》,第 85 页。

6. 参见蓓丝·罗德:《哲学之外的斯宾诺莎》,爱丁堡:爱丁堡大学出版社,2012年,第 1 页。

6 思想支脉

要点 ⚷

- 斯宾诺莎的论述从非常抽象的形而上学的观点出发，最后落在一个"人如何过上美好的生活"这样的具体论断上。对知识和情感的本质的讨论使他做出了这样的论述。

- 斯宾诺莎在他的认识论、心理学和行动理论中阐述了一些他最具影响力、可能也最引人争议的学说，比如他的决定论。

- 虽然斯宾诺莎的形而上学引发了大量关注，但他关于伦理学的具体观点有时却被忽视了。

其他思想

在《伦理学》中，本尼迪克特·德·斯宾诺莎的形而上学（研究存在的本质的一个哲学分支）和他的认识论（对知识的本质进行的哲学探究）相互交织。后者主要是他关于身心之间平行关系的看法的一种延伸。斯宾诺莎认为，一个个体的心灵不是存放个人精神内容的、与外界相隔绝的仓库，它自己就是与身体紧密相关的世界的一部分；某种程度上来说，身心是一枚硬币的两面。知识是智力的关键组成部分。斯宾诺莎对通过感觉获得的知识（不完备的观念）与通过理性或是智力直觉所获得的知识（完备的观念）进行了区分。我们的感觉和想象力给予我们的是"令人困惑和残缺的"[1]关于物质事物的精神画面，这些画面是片面且主观的。相反，斯宾诺莎认为理性和智力直觉（智性直观*）会让我们认识永恒的真理。强调理性的斯宾诺莎似乎在否认同时代的约翰·洛克[*2] 提出的经验主义，后者认为所有的知识都是通过感觉获得的，否认纯粹通过理性或者智力直觉获得的知识的存在。于是斯宾诺莎将自己归入了理性主义者的阵营（但是他与经验主义之间的关系仍是学者们争辩的话题）。

斯宾诺莎将其认识论用于他的情感心理学和行动理论上。后者的核心是欲求原则："每一个自在的事物莫不努力保持其存在。"[3] 斯宾诺莎在对情感或"情状"的讨论中陈述过这个原则。他认为情状源自快乐、悲伤和欲望——三种心灵活动的基本形式。快乐是生存力量增强的结果，而悲伤恰恰相反；欲望是对保持存在所做的努力本身（即保持生存的努力）。所有其他情状都由这些因素的组合形成。例如，斯宾诺莎说："希望是一种不稳定的快乐，这种快乐出现于我们对过去或未来某一事件的结果并不确定的时候。"[4]

斯宾诺莎区别了被动的情状和主动的情状。喜悦和悲伤被认为是被动的情状，因为它们是"不完备的观念"，是心灵被外界因素影响的方式。然而一旦心灵清楚地领会了喜悦和悲伤之后，它们就会被转化为主动的情状。[5] 这是因为人通过智力直觉明白了这些情状是遵循决定论法则一步一步形成的。

于是我们被引向了斯宾诺莎的行动理论，它基于这样的论点：一个人先在理智上承认自由意志是一种假象，然后才会采取正确的行动。对斯宾诺莎而言，只有当我们意识到行动是由一系列因果关系促成的，我们才会得到自由。换句话说，自由不是意志的自由，那是一种假象。我们关于善恶的语言是建立在自由意志的错误概念上的，斯宾诺莎的《伦理学》基于欲求的原则对其做了修正。通过把情状从被动转为主动并且理解其本质，心灵才能努力地求生。

> "我所理解的神是一个绝对无限的存在，即一个拥有无限属性的实体，其中每一个属性都显示了一种永恒、无限的本质。"
>
> —— 本尼迪克特·德·斯宾诺莎：《伦理学》

思想探究

帮助我们从《伦理学》开篇抽象的形而上学论断过渡到具体的心理学和伦理学论断的是智力直觉（直觉知识）这一基本观点。智力直觉是除了理性和通过感觉获得的知识之外的第三种知识。斯宾诺莎模糊地将其定义为"由对神的某一属性的形式本质的完备观念出发，进而获得对事物的形式本质的完备知识"。[6]斯宾诺莎所称的智力直觉是指：要理解或者直觉认识到，一个事物是必然遵循单一实体的本质的。这既不是通过理性获得的知识，也不是通过感觉获得的知识，而是另外一种直接的直觉知识。换言之，只有进入认识的状态，世界的决定论本质才能被理解，直觉知识在其中起到了关键作用。

斯宾诺莎关于智力直觉的论述有赖于他的形而上学体系里早期的重要内容，即区分能动的自然*与被动的自然*。这区分了产生自然的"主动"原则和它的"被动"产物。能动的自然是原因，而被动的自然是必然产生的效应。智力直觉成了能够理智地、直接地了解自然整体（虽然只是被动的自然）的一种方法。智力直觉把握的是被动的产物，而不是自然的主动原因，它让我们认识到，每一个受到研究的具体事物必然来自神的本质——除此之外没有其他可能。这种认识事物的方式是从被动的情状里解放出来的基础。

被忽视之处

颇为讽刺的是，《伦理学》中最为人忽视的部分就是伦理学的内容。这本书最初遭遇到严厉批评，斯宾诺莎被指控为宣扬无神论 *，他的观念被全欧洲否定。当时人们主要关注的是他关于神的观念。即便是在该书出版一个世纪之后的 18 世纪 80 年代，德国泛神论争议使《伦理学》受到更多批评讨论的时候，第一部分和第二部分中关于神和自然的论断还是争论的焦点，而不是第三部分到第五部分的伦理学内容。[7] 这种关注的后果是斯宾诺莎的学说被解读为遵循泛神论传统的形而上学，[8] 这个传统是正统宗教所不能接受的。但一种较为宽容的解读是斯宾诺莎力求完全脱离传统，他的论述方向是如何理解人类自由。过分关注作为某种传统的形而上学而无视书中关于伦理学的洞见，这是不公正的。因为有一层意义被隐没了，那就是书中的每一部分都意在指引读者去了解斯宾诺莎关于"一个人如何过上美好生活"的批判性探究，斯宾诺莎把这种生活看作是有福佑的生活。

---

1. 本尼迪克特·德·斯宾诺莎：《伦理学》，选自《斯宾诺莎读本:〈伦理学〉及其他著作》，埃德温·柯利译，普林斯顿:普林斯顿大学出版社，1996 年，第135 页。
2. 参见亚历山大·道格拉斯:"斯宾诺莎是自然主义者吗?"，《太平洋哲学季刊》第 96 卷，2005 年第 1 期，第 85 页。
3. 斯宾诺莎:《伦理学》，第 159 页。

4. 斯宾诺莎:《伦理学》,第 190 页。

5. 斯宾诺莎:《伦理学》,第 247 页。

6. 斯宾诺莎:《伦理学》,第 141 页。

7. 参见弗里德里希·海因里希·雅各比和杰拉德·瓦雷:《莱辛和雅各比关于斯宾诺莎的对话:后世争论的摘录文本》,美国大学出版社,1988 年。

8. 参见约翰·杜威:"斯宾诺莎的泛神论",《思辨哲学学刊》第 16 卷,1882 年第 3 期,第 249—257 页;F. C. S. J. 科普尔斯顿:"斯宾诺莎与德国唯心主义哲学家思想中的泛神论",《哲学》第 21 卷,1946 年第 78 期,第 42—56 页。

7 历史成就

要 点 ⚷━━

- 斯宾诺莎的《伦理学》使用几何学的风格建构了一个复杂的论证系统，这个系统内部联系紧密。虽然这种风格非常优雅，但它的缺点是如果其中一个论证或定义被否定了，那么整个作品也就解体了。

- 《伦理学》在今天被视为最重要的现代哲学作品之一，影响了许多伟大作家。但这部作品出版时却被认为是否认上帝存在的，它成了 17 世纪的重大丑闻。

- 虽然《伦理学》的结构是相互紧密联结的，似乎只要其中任何一个部分被否定，整个作品就会解体，但是它依然极富创新色彩，并且它包含着关于人类生活的敏锐观察。

观点评价

本尼迪克特·德·斯宾诺莎在《伦理学》中的宏大任务是通过给出一系列抽象的定理（不证自明的论断）和定义，再通过一系列的论证，最终呈现神、人类知识、情状的心理机制和自由的全貌。这项任务本身涉及的范畴之广以及斯宾诺莎在执行这项任务时用心之精密，都会吓退读者。每一个新论证都是在之前的定义、定理和论证的基础上提出的，全文自成体系。有时这本书会让人觉得是座迷宫：斯宾诺莎在写作时采用几何学的风格，频繁地参照之前的论证。

斯宾诺莎采用几何学的方法所创造出来的这一系统有一个巨大的缺点：如果一个人否定了其中的某个证据、定理或定义，所有其他部分都将随之被否定。这种论证结构（即几何学的方法）使这本书宏大的系统性成就会因为一些细节遭否定就被抹除，系统的整体性亦被忽视，斯宾诺莎的很多观点都可以被一些充分的理由推翻。例如，他颇具野心地试图去证明上帝的存在。虽然证明上帝的存在是当时的哲学领域中非常重要的一部分，但是哲学史后来的发展却质疑了这些做法。人们对斯宾诺莎论证的内部逻辑也有进一步的疑问，因为他们并不清楚是否接受斯宾诺莎的所有定义和定理就意味着必须接受他的所有论证。因为有些观点不能被前文中的定义或定理证明，斯宾诺莎在引出这些观点时就会依赖文中的评注或新的定义。

斯宾诺莎的结论往往要比得出这些结论的形式推理更为有趣，这些结论也值得当代哲学家重新修正。

> "你要么是个斯宾诺莎的信徒，要么就根本不是一个哲学家。"
>
> ——格奥尔格·威廉·弗里德里希·黑格尔:《哲学史讲演录》

当时的成就

因为这本书晦涩难懂，所以它较为引人争议的观点也可能被误读，比如关于上帝的本质那些观点。在出版之后的大约一个世纪里，《伦理学》的神学观念都被解读为无神论的，即否认上帝的存在。但后来的解读者却认为它是泛神论的，即认为上帝无处不在（一种与大多数一神论的宗教相冲突的信仰）。

斯宾诺莎关于宗教的思想被认为是激进的。这意味着他的观念在哲学领域中被视为异端；一个人要是被指控为支持"斯宾诺莎主义"就是一种侮辱，是要为自己辩护的。然而这些对他思想的攻击来自一种文化上的偏见，而不是来自对他的哲学论点的公正分析。在"上帝／神"这个词后面加上"或自然"这个说法，在他写作的历史环境里也不可能作为常态被接受。同样，斯宾诺莎反对神人同性的概念，这也让许多哲学家感到恼火。

《伦理学》依然是最重要的哲学著作之一，因为它的讨论范围涉及各种话题，从存在的本质到心灵和身体之间的关系以及在一个决定论的世界里获得幸福的方式。这些话题已经超出了哲学的范围，延伸到了认知疗法（一种心理治疗方法）、环保主义、政治学和艺术的领域。例如，阿根廷作家豪尔赫·路易斯·博尔赫斯*的作品将文学、幻想和哲学熔为一炉，他就承认受斯宾诺莎影响极大。[1] 同样，18世纪和19世纪早期最著名的德国作家约翰·沃尔夫冈·冯·歌德*盛赞斯宾诺莎，因为后者的著作让他年轻时从过度的激情中沉静下来。[2] 另外，斯宾诺莎的作品还被认为是弥合了东西方思想分歧的桥梁。

《伦理学》得到如此普遍的重视的另一个原因可能是它的写作风格。斯宾诺莎放弃了容易理解的写作方式，有意地使用数学术语来准确地表达他的观点，同时，这种写作方式能使本书在所有文化中进行有意义的传播。当然，斯宾诺莎并没有完成这个目标，《伦理学》中的语言反映了这本书受到的不同历史影响，其中包括斯宾诺莎的马拉诺犹太教思想背景、勒内·笛卡尔在启蒙运动*中宣扬的理性主义，甚至这种数学写作风格本身也源自欧几里得的《几何原本》。

局限性

斯宾诺莎沿用的是欧几里得几何学的写作风格——以这个命题

为例：两条无限延伸的平行线永远不会相交，这用直觉判断显而易见。另外一个类似的逻辑原理是同一律（A=A）。虽然这种原理对所有的理性存在来说是不证自明的，但是它本身却不能被证明。问题在于，是否存在任何自然而然的理由让读者接受斯宾诺莎的基本命题。和直观的逻辑原理或几何公理不同，斯宾诺莎的出发点复杂又模糊，接受它们似乎要多花点力气。这样，抽象定义和相对更具体的论断（这就是该书五个部分的结构方式）之间的平衡可能要被打破了。[3] 因为，既然对较具体的论断判断真伪依赖的是较抽象的论断，那么否认那些抽象部分也就是否认那些它们自己推导出的具体部分。另外，既然本书中的抽象论断在某种程度上依赖于较具体的论断，那么这种从抽象到具体的几何学方法就会遭到破坏。尽管有如此种种不足，《伦理学》还是在哲学史上作出了创新性的贡献，并发展出了一套极为独特的、对人类生活有细致观察的关于存在的理论。

1. 参见马塞洛·阿巴迪："博尔赫斯镜中的斯宾诺莎"，《斯宾诺莎研究：国际与跨学科系列》第 5 卷，1989 年，第 29—42 页。

2. 斯图尔特·汉普什尔：《斯宾诺莎》，曼彻斯特：曼彻斯特大学出版社，1956 年，第 18 页。

3. 参见古托姆·弗洛施塔特："斯宾诺莎的知识论"，《探究》第 12 卷，1969 年第 1—4 期，第 41—65 页。

8 著作地位

要点 ⚿

- 通过斯宾诺莎的早期作品以及现存的他同别人的通信，我们可以深入地了解他的思想发展过程。
- 《伦理学》解答了许多笛卡尔及其追随者所提出的问题，比如心灵和身体的关系。在推广对《圣经》的学术研究的过程中，斯宾诺莎强有力的写作风格起到了重要作用。
- 因为被革出教门和一则写于他死后不久的生平词条，斯宾诺莎在死后的一个世纪里声名狼藉。

定位

《伦理学》是本尼迪克特·德·斯宾诺莎最伟大的作品。他确实也写过许多其他的作品，从这些作品中我们可以进一步了解《伦理学》，这与我们了解这些作品本身所包含的观点是同样有用的。这些作品展现了斯宾诺莎在被逐出教会和离开人世之间的20年中思想逐渐发展、趋于一致的过程。在这些作品中最值得注意的是斯宾诺莎与一系列人物的通信和《知性改进论》（1662）、《笛卡尔哲学原理：依几何学方式证明》（1663）、《神学政治论》（1670）以及未完成的《政治论》。读者最好按照这个顺序阅读。

现在大约有50封斯宾诺莎的个人书信尚存于世，[1]这些书信的写作时间大概是从1660年直到斯宾诺莎1677年去世。它们除了展现作者的人格个性以外，还透露出早在1660年斯宾诺莎就开始深入思考几个关键的问题，比如说上帝和理性的统一，这个论点后来出现在《伦理学》里。

《知性改进论》("改进"的意思是"修改"或"改正")是斯宾诺莎早期创作中一部重要的作品，斯宾诺莎在 17 世纪 50 年代晚期开始写作此书，在书中他提出了获得真正理解的正确方法。[2] 和书信一样，这部作品也力图把清楚的、独特的真实观点同不足的、误导我们的观点区分开来。这本书证明了勒内·笛卡尔给斯宾诺莎带来的显著影响。在写于 1663 年的《笛卡尔哲学原理：依几何学方式证明》一书中，笛卡尔对斯宾诺莎的影响最为明显。[3] 这本书是应一些想更好地理解笛卡尔思想的朋友之求所写，它的重要之处在于标志着斯宾诺莎开始向《伦理学》的几何学写作风格转变，这也是唯一一部斯宾诺莎实名出版的作品。

写作时间与《伦理学》最接近的一部作品是出版于 1670 年的《神学政治论》。[4] 这本书用一种平实的语言介绍了后来《伦理学》用几何学的语言所表达的观点。《神学政治论》为宗教自由提供了理性主义的基础，它是在斯宾诺莎创作《伦理学》的间隙出版的。

《政治论》是斯宾诺莎的最后一部作品，一直尚未完成甚至仍有一点神秘。[5] 斯宾诺莎在 1676 年年中完成《伦理学》后开始写作这本书，斯宾诺莎的本意是把它当作《神学政治论》的续作。斯宾诺莎写作这本书的目的是为了说明不同宪法下的各个国家如何能良好地运转，于是这本书也起了支持民主制的作用。这些作品虽然也有人注意，但是斯宾诺莎的声誉大部分还是来自他的主要作品《伦理学》。

> "《伦理学》不仅是斯宾诺莎的杰作，还是他毕生的事业。我们从他的通信中得知，斯宾诺莎于 17 世纪 60 年代开始写作此书，1665 年时这部作品的草稿已经基本成形，然而斯宾诺莎将其放在一边，开始写作《神学政治论》，后者于 1670 年问世。斯宾诺莎出版关于笛卡尔哲学的论述就是为了给《伦理学》做准备，而写作《神学政治论》也是出于相似的动机。"
>
> ——埃德温·克利：《斯宾诺莎的人生与哲学》

整合

如果斯宾诺莎不出版《伦理学》的话，他的整体创作看起来就会前后不太一致。从《伦理学》之前的作品中所关注的话题可以看出，斯宾诺莎对笛卡尔及其追随者的哲学有明显的兴趣。在斯宾诺莎写作时，笛卡尔哲学刚刚开始在荷兰的哲学论辩中占据主导地位。另一方面，斯宾诺莎似乎相当关心神学和政治学的问题，他一手促进了《圣经》的世俗研究和解读。这两种研究兴趣就是他生前出版的最重要的两部作品的主题，即《笛卡尔哲学原理》和《神学政治论》。

然而，《伦理学》以某种方式将上述两部作品综合起来了。例如，在《笛卡尔哲学原理》中，斯宾诺莎在写作哲学内容时引入了几何学的方法。这本书同时也为《伦理学》中提到的笛卡尔式的问题提供了背景，这在很大程度上是对笛卡尔及其追随者提出的问题（比如心灵和身体的关系）的一种回应。《神学政治论》关注的却是《伦理学》研究过的主题，比如对神人同性（即神有人的特性）的否认、情状的心理学机制和在宗教中的地位以及自由的概念等。

意义

《伦理学》把斯宾诺莎之前关心的问题系统性地综合起来，采用几何学的写作风格来论证观点。它涉及了许多领域，从形而上学、神学到认识论、心理学，这些领域之间是互相关联的。

虽然《伦理学》是斯宾诺莎最重要的作品，但在它出版之后的一个世纪里，让斯宾诺莎以思想家而闻名的还有其他因素。斯宾诺莎之所以声名不佳，是因为人们散布关于他生平的谣言，特别是皮埃尔·培尔 * 在《历史批判辞典》（1697）中为他所写的传记。斯宾诺莎早年被逐出教会，《神学政治论》的出版伴随着不少争议，他又和共和主义的领导人扬·德·维特有交往，这都让不怀好意的人有理由把《伦理学》看成对无神论的辩护。斯宾诺莎认为存在的最高境界是对上帝的"理智的爱"（"对神圣智性的爱"），这让上述对斯宾诺莎的理解显得过于严苛。直到他逝世后大约一个世纪，在德国知识分子开始所谓的泛神论争辩之后，大家才调整了对斯宾诺莎的评价。这说明对《伦理学》的解读有了新的进步，它不再被视为对无神论的支持，而是对一种泛神论（神即自然）形式的发扬。虽然这种看法已经比较宽容了，但斯宾诺莎思想的当代阐释者们仍然在争辩：我们应该把他看成是一个无神论者、泛神论者还是一个万有在神论 * 者（认为上帝存在于所有自然中）？[6]

---

1. 本尼迪克特·德·斯宾诺莎：《斯宾诺莎通信集》，纽约：罗素兄弟出版社，1928 年。

2. 本尼迪克特·德·斯宾诺莎："知性改进论"，《伦理学：知性改进论与书信选集》，塞缪尔·舍利译，印第安纳波利斯：哈克特出版社，1998 年。

3. 参见本尼迪克特·德·斯宾诺莎：《笛卡尔哲学原理：附录形而上学思想》，塞缪尔·舍利译，印第安纳波利斯：哈克特出版社，1998 年。

4. 本尼迪克特·德·斯宾诺莎：《神学政治论》，剑桥：剑桥大学出版社，2007 年。

5. 本尼迪克特·德·斯宾诺莎："政治论"，《神学政治论与政治论》，弗朗西斯科·科尔达斯科编，纽约：库里尔出版集团，2013 年，第 267—388 页。

6. 参见吉纳维芙·劳埃德：《劳特利奇哲学导论·斯宾诺莎与〈伦理学〉》，伦敦：劳特利奇出版社，1996 年，第 40 页。

第三部分：学术影响

9 最初反响

要点 🔑

- 《伦理学》被想当然地认为是支持无神论的、违背道德的，因此它最初遭到了盲目的谴责。一个世纪后，哲学家们才开始在这部作品中看到丰富的智慧。

- 虽然著名哲学家莱布尼茨在私底下和斯宾诺莎有联系并采用了他的一些观点，但是他曾公开参与对斯宾诺莎的强烈谴责。

- 在斯宾诺莎逝世一百年后的泛神论争辩中，著名哲学家戈特霍尔德·莱辛 * 承认斯宾诺莎的泛神论影响了他，这使斯宾诺莎的作品逐渐得到读者接受。

批评

在本尼迪克特·德·斯宾诺莎的《伦理学》出版后的一段时间里，它得到的反应不佳，这很大程度上要归咎于皮埃尔·培尔在他读者众多的《历史批判辞典》（1695）中指责《伦理学》支持无神论、违背道德。思想史学家彼得·盖伊 * 说，培尔的这部辞典"误导了读者们一个世纪"。[1]

在 18 世纪 80 年代的德国，人们对斯宾诺莎的态度第一次出现明显缓和。这次缓和源自戈特霍尔德·莱辛和弗里德里希·亨利希·雅科比 * 之间的争论，也就是后来人们所知的泛神论论争。莱辛在寻找有别于当时德国启蒙运动主流的思想形式时，坦承自己在斯宾诺莎的《伦理学》中找到了可以认同的体系。虽然这番坦承带来的反响对莱辛不是十分有利，但它引起的讨论却大大扭转了大家对斯宾诺莎的看法。与培尔的《历史批判辞典》中被夸张、被讽刺的无神论者形象不同，斯宾诺莎开始被视为一个逻辑自洽的形而上学 / 神学体系的开创者，这个体系就是泛神论。不久之后，参与了18世纪晚期和19世纪早期的德国唯心主义运动 * 的哲学家都能在《伦理学》中找到吸引自己的部分，其中最引人注意的要属《伦理学》中形而上学和神学的内容（第一部分和第二部分），而不是伦理学的内容（第三部分到第五部分）。例如著名的德国哲学家格奥尔格·威廉·弗里德里希·黑格尔在他的《哲学史讲演录》（1837）中提到，他认为斯宾诺莎关于实体的看法是"所有真实观点的基础"，而斯宾诺莎的观点被苛责的真正原因是，那些批评家一想到自然的同一性会把自己消灭就受不了了。[2] 在很大程度上，泛神论论争开启了一种关于《伦理学》在西方哲学中的地位的讨论模式。这种讨论模式是：大家先是把《伦理学》看成是一些其他论题的代表作，后来只关注它前两部分关于神和自然的内容。这样做的后果就是它后三个部分的内容被完全忽视了。

> "他袒护人类能做出的最无耻、最残暴、最肆无忌惮的行径，这比惦记着异教神明的诗人要荒谬百倍。他要么是没看见，要么是看见了，但是固执己见，以致刚愎自用。一位贤者应该使出浑身解数开出一片新土来，而不是如此虚伪、荒唐、可怕地去犁田。"
>
> ——皮埃尔·培尔：《历史批判辞典》

回应

《伦理学》出版于斯宾诺莎去世之后，他不可能对这部作品的批评者做出回应了。但是斯宾诺莎留下的书信证明，即便是同侪的善意批评，他都不屑一顾。³事实上，有人会和他通信，讨论一下他作品中的难点，但无论何时迎面遇上这个人，他经常不理不睬；在他晚年时，这种情况越来越多。书信里透露的这一倾向让人有理由相信，因为他不慌不忙的性格，即使需要面对《伦理学》招致的批评，他也能够保持平静。不管是被逐出教门还是17世纪70年代因《神学政治论》而起的纷纷扰扰，斯宾诺莎面对它们时多多少少都能够控制自己。《神学政治论》是斯宾诺莎的所有作品中，如果不是在形式上也是在内容上和《伦理学》最相像的。当然，这只是一种推测而已，就好像认为斯宾诺莎没写完的《政治论》会回应对《伦理学》的批评一样，那也只是一种期望罢了。但是鉴于斯宾诺莎成年以后思想就再无大变，想要说服他肯定不是一件容易的事。

哲学史上的一段有趣的插曲会让我们更加清楚斯宾诺莎对待批评的态度。著名的学者、哲学家戈特弗里德·威廉·冯·莱布尼茨通常被认为是理性主义传统的代表人物。这项传统开始于勒内·笛卡尔，经由斯宾诺莎，结束于莱布尼茨。莱布尼茨曾经和斯宾诺莎通过信，甚至在1676年《伦理学》已经写完但尚未付梓时拜访过斯宾诺莎。但由于斯宾诺莎声名不佳，莱布尼茨对这次拜访秘而不宣。在生前出版的作品中，莱布尼茨对斯宾诺莎都持批评态度，他还特意将自己坚持自由意志存在的观点和斯宾诺莎的决定论相比较。但他确实将斯宾诺莎的不少观点融入了自己的作品，他死后出版的作品证明，私底下他和斯宾诺莎一样相信决定论的观点。⁴在18世纪哲学的语境中，任何观点只要和斯宾诺莎相关就被打入冷宫，其命运由此可见一斑。

冲突与共识

　　在整个 18 世纪，哲学家们都在气势汹汹地谴责斯宾诺莎的作品。他们多半是受法国哲学家皮埃尔·培尔所写的传记的影响，把斯宾诺莎的作品草率地、不公正地判定为一部无神论者的作品，何况这个人曾被阿姆斯特丹的犹太教会谴责为异端 *，他的《神学政治论》还被荷兰归正会宗教议会（即总议会）查禁。这就是说，几乎没有学者用公正的眼光看待斯宾诺莎的作品。一个值得注意的例外是在莱布尼茨的私人信件中发现的。但是，莱布尼茨终其一生都对这件事只字不提；虽然莱布尼茨私底下持决定论的世界观，但在公开场合他却抨击斯宾诺莎式的决定论，把这种决定论和自己的形而上学相对照，认为后者捍卫了自由意志。

　　可能因为斯宾诺莎比哲学史上任何一位思想家都激进，在遭受了一个世纪的蔑视之后，他的名誉才开始得到恢复。斯宾诺莎被看作无神论者，再加上他自己激进的、决定论的观点，使其他哲学家普遍把斯宾诺莎主义看作是一种哲学的罪恶。直到斯宾诺莎去世大约一个世纪后发生的泛神论论争中，哲学家们才敢坦承斯宾诺莎曾经影响了自己。

---

1. 彼得·盖伊：《启蒙：文选汇编》，纽约：西蒙与舒斯特出版社，1973 年，第 293 页。

2. 吉纳维芙·劳埃德：《劳特利奇哲学导论·斯宾诺莎与〈伦理学〉》，伦敦：劳特利奇出版社，1996 年，第 16 页。

3. 参见本尼迪克特·德·斯宾诺莎："异议与回答"，选自《斯宾诺莎读本：〈伦理学〉及其他著作》，埃德温·柯利译，普林斯顿：普林斯顿大学出版社，1996年，第 146 页。

4. 参见伯特兰·罗素：《西方哲学史》，牛津：劳特利奇出版社，2004 年。

10 后续争议

要点 ⚷━┉

- 19 世纪更加自由的社会风气让哲学家们对斯宾诺莎的态度由否定转为盛赞。后来，其他领域伟大的思想家，比如西格蒙德·弗洛伊德*和阿尔伯特·爱因斯坦*都曾经说，斯宾诺莎对他们的影响非常深远。

- 如果"正统"的斯宾诺莎主义者意味着要接受他的所有观点，那么几乎没有学者是正统的斯宾诺莎主义者。但是今天很多不同领域学者的作品深受斯宾诺莎的影响，这些学者可以被称为斯宾诺莎主义者。

- 对斯宾诺莎《伦理学》的研究仍在继续。这本书不仅对现代形而上学、认识论、心理学和伦理学贡献良多，还和心灵哲学、激进政治思想、动物行为学*和心理分析等领域的当代论争有关。

应用与问题

在本尼迪克特·德·斯宾诺莎的《伦理学》出版后将近三个半世纪里，针对这部作品的普遍抵制逐渐被广泛的赞誉取而代之。促成这种转变的是在宗教问题上日益削弱的共识、政治的自由化以及哲学兴趣的转向。随着 18 世纪 80 年代泛神论论争在德国展开，对斯宾诺莎几乎一边倒的否定开始被颠覆了。这场论争开始的标志就是戈特霍尔德·莱辛承认自己是斯宾诺莎的信徒。接下来的争论引发了第一波对斯宾诺莎的正面评价。格奥尔格·威廉·弗里德里希·黑格尔甚至断言阅读所有的哲学著作都要从阅读斯宾诺莎开始。[1]

斯宾诺莎研究者皮埃尔-弗朗索瓦·莫罗*认为,泛神论论争标志着启蒙运动的结束,推动了浪漫主义*这一文艺运动的发展,并让大家开始尊崇斯宾诺莎的思想。[2] 不管莫罗的看法有无其道理,到了19世纪,哲学界的舆论开始向欣然接受《伦理学》的方向转变了。例如,人们对无神论的态度有很大的转变,弗里德里希·尼采*就认为斯宾诺莎是促成这种转变的先驱。

20世纪,西格蒙德·弗洛伊德高度评价斯宾诺莎,伯特兰·罗素*也多次提及斯宾诺莎的伦理学观点,阿尔伯特·爱因斯坦对斯宾诺莎也赞赏有加。当爱因斯坦受到一位犹太教士的追问时,他声称自己信奉的是斯宾诺莎的上帝,这种上帝是理性的,不关心人间事。现在大家对斯宾诺莎的兴趣较以往更为浓厚,有一批历史、文本和哲学层面的研究成果正在迅速涌现。

> "斯宾诺莎是哲学家的救世主,而最伟大的哲学家们,不管是对他的神秘学说敬而远之还是趋之若鹜,都只是他的使徒罢了。"
>
> —— 吉尔·德勒兹和菲力克斯·迦塔利*:《什么是哲学?》

思想流派

当代有一批哲学家的研究框架可以被称作斯宾诺莎主义。像埃德温·克利*和乔纳森·班内特*这样的思想家都以斯宾诺莎的观点作为自己研究的来源,他们自己的研究着重于严谨缜密地发展心灵哲学和进行逻辑分析。克利把斯宾诺莎关于思想和广延的关系的分析看作上帝的特性,[3] 班内特在他的《斯宾诺莎〈伦理学〉研究》中比较了当今许多心灵哲学和形而上学之间的最新论争。[4]

20世纪法国哲学家吉尔·德勒兹*把认知疗法与激进政治学结合，用以证明阅读斯宾诺莎有助于读者在面对现今的社会和政治问题时打开新思维。[5]这些看法都为德勒兹新创立的动物行为学提供了依据。动物行为学是从生物学的角度研究人类和社会行为的学科。

一组人数少但规模仍在扩大的评论家提出了与德勒兹的分析比较类似的看法：一个人只有在更全面、更完整的历史语境中才能理解斯宾诺莎。[6]罗伯特·S.科灵顿*就把斯宾诺莎的观点，特别是对"被动的自然"和"能动的自然"的区分，应用到他所称的"狂喜自然主义"中。这种说法把斯宾诺莎的自然主义*和德国唯心主义哲学家弗里德里希·谢林*对"自然的自发性"的强调结合起来，其中还有一些实用主义哲学传统的因素，用以解释自然中的神圣经验。[7]有些斯宾诺莎的信徒并不完全是哲学家，例如约翰·沃尔夫冈·冯·歌德、[8]豪尔赫·路易斯·博尔赫斯[9]以及西格蒙德·弗洛伊德。[10]但需要我们注意的是，实际上这些人都没有把斯宾诺莎的作品整体性地应用到任何课题中去。

当代研究

虽然斯宾诺莎的重要性得到了一致认可，但对于一些人来说，他的思想起到了不同的，甚至是完全相反的作用。例如，德勒兹不顾争议，把斯宾诺莎归于经验主义的旗下。[11]在德勒兹的观点中，这表示否认超验实体的存在，因为斯宾诺莎将上帝/神等同于自然。斯宾诺莎认为心灵和身体同样受决定论的支配，对于德勒兹来说，这一理解有某种政治含义：这种说法否定了笛卡尔坚持的主体性的独立本质，说明了心灵是世界的一部分，它与决定其状态的种种关系紧密相连。德勒兹解读斯宾诺莎的方法集中在其哲学的实用意义上，他认为斯宾诺莎把本体论这种理论学科同伦理学这种实用学科结合起来了。

但如今许多别的思想家恰恰相反，只强调《伦理学》中较为理论化的内容。[12] 例如，关于这本书中的逻辑和心理学的关系，他们进行了一场公开辩论。班内特认为斯宾诺莎没有恰当地区分逻辑和心理学，[13] 阿尔伯特·巴尔兹*认为斯宾诺莎对逻辑的论述完全把心理学排除在外了。[14] 另一个尚在进行中的讨论是关于斯宾诺莎的因果律概念，心灵和身体的因果关系这一问题就与这个概念有关。查尔斯·贾赫特说《伦理学》认为没有一种心灵事件可以导致身体事件，反之亦然。[15] 但是像唐纳德·戴维森和奥利·柯艾斯蒂南这样的研究者就不同意这种说法。[16] 他们认为，只要了解一下他们提出的"透明"因果律，就知道斯宾诺莎否认因果关系的存在是站不住脚的。在"透明"因果律中，如果 X 导致 Y，X 等同于 Z，那么 Z 就导致 Y。所以，你脑中的生理变化导致你的胳膊肌肉收缩了，这种脑中变化是一种精神决定，那么这种精神决定就导致了你的肌肉收缩。

1. 参见格奥尔格·威廉·弗里德里希·黑格尔：《哲学史讲演录·第三卷：中世纪与近代哲学》，E. S. 霍尔丹和弗朗西斯·H. 西姆森译，林肯：内布拉斯加大学出版社，1995 年，第 2 部分，第 1 章，A，第 2 页。

2. 皮埃尔-弗朗索瓦·莫罗：《斯宾诺莎：经验与永恒》，巴黎：法兰西大学出版社，1994 年，第 420—421 页。

3. 埃德温·柯利：《斯宾诺莎的形而上学：阐释与论述》，马萨诸塞州坎布里奇：哈佛大学出版社，1969 年。

4. 乔纳森·班内特：《斯宾诺莎的〈伦理学〉研究》，纽约：哈克特出版社，1984 年。

5. 吉尔·德勒兹：《斯宾诺莎与表现问题》，马萨诸塞州坎布里奇：麻省理工学院出版社，1990 年。

6. 参见威利·哥特谢尔：《斯宾诺莎的现代性：门德尔松、莱辛与海因》，威斯康星州麦迪逊：威斯康星大学出版社，2004 年。

7. 罗伯特·S.科林顿：《狂热的自然主义者：世界的迹象》，印第安纳波利斯：印第安纳大学出版社，1994 年。

8. 斯图尔特·汉布什尔：《斯宾诺莎》，曼彻斯特：曼彻斯特大学出版社，1956 年，第 18 页。

9. 参见马塞洛·阿巴迪："博尔赫斯镜中的斯宾诺莎"，《斯宾诺莎研究：国际与跨学科系列》第 5 卷，1989 年，第 29—42 页。

10. 参见沃尔特·伯纳德："弗洛伊德与斯宾诺莎"，《精神病学》第 9 卷，1946 年第 2 期，第 99—108 页。

11. 吉尔·德勒兹：《斯宾诺莎与表现问题》，马萨诸塞州坎布里奇：麻省理工学院出版社，1990 年；吉尔·德勒兹：《斯宾诺莎的实践哲学》，罗伯特·赫利译，旧金山：城市之光出版社，1988 年。

12. 参见乔纳森·班内特：《斯宾诺莎的〈伦理学〉研究》，纽约：哈克特出版社，1984 年。

13. 乔纳森·班内特：《斯宾诺莎的〈伦理学〉研究》。

14. 阿尔伯特·G.A.巴尔兹：《霍布斯与斯宾诺莎哲学中的观念和本质》，纽约：哥伦比亚大学出版社，1918 年。

15. 查尔斯·E.贾勒特：《斯宾诺莎入门导读》，纽约：连续体出版社，2007 年。

16. 唐纳德·戴维森："斯宾诺莎关于情动的因果论"，《欲望与情动：作为心理学家的斯宾诺莎》，伊尔米亚胡·约维尔编，莱顿：布里尔出版社，1999 年，第 95—111 页；奥利·科伊斯汀："因果关系、意向性和同一性：心身互动"，《比率》第 9 卷，1996 年第 1 期，第 23—38 页。

11 当代印迹

要点 🔑

- 自从 18 世纪 80 年代的泛神论论争使哲学家可以讨论《伦理学》的多方面贡献之后，这本书就在各种学科领域激起了思想的火花。
- 斯宾诺莎写作《伦理学》的目标是质疑目的论思维（研究事物的目的）、自由意志的观念以及上帝与世界相分离的观点。即便在今天，这本书依然颇为激进。
- 近来对斯宾诺莎的哲学研究不仅关注他的形而上学和心灵哲学内容，也关注他的政治学甚至生态思考。

地位

本尼迪克特·德·斯宾诺莎的《伦理学》自首次出版后便一直有其重要价值。在过去的三个世纪里，随着哲学家们的关注点慢慢向接受斯宾诺莎论点的方向转变，斯宾诺莎也变得越来越重要。在有些领域确实如此，例如，现代科学就致力于发现既适用于自然又适用于人类的法则。政治学中对目的论思维和等级制度进行激烈批判也是一例。总之，许多领域的思想家都在《伦理学》中发现了宝贵的灵感来源。

许多其他领域的研究者也对这本书产生了兴趣。[1] 斯宾诺莎一直在努力攻克形而上学、逻辑学或心灵哲学中的问题，有些研究者就会关注这些问题在当下的意义。[2] 还有一些研究者颇为推崇斯宾诺莎思想中的政治内涵、他探求思想界限的努力以及他的思想给生态学带来的积极影响——否定人类在自然中的特殊地位。[3]

斯宾诺莎能时来运转，也是因为人们的宗教观念改变了。在《伦理学》出版后的一个世纪里，这本书被指控为支持无神论，因而不被大众所接受。在那个年代里，公开表达无神论观点可能会招致杀身之祸。但在18世纪80年代，德国思想家之间的泛神论论争为斯宾诺莎的思想扩大了接受面以后，人们对这部作品的反应有了明显的改变。到19世纪晚期，像弗里德里希·尼采这样反教权（即反对神职人员拥有权力）的思想家，可以在否认自由意志或驳斥"善恶的二元对立存在于人类本质中"时，公开承认斯宾诺莎是自己的灵感来源。尼采在毕生的所有作品里处处提及斯宾诺莎，有些是盛赞他是自己的向导，有些则是对他的批评。[4] 同样，现代科学阵营中把自己视为自然主义者的人们也认为，斯宾诺莎支持自然主义（自然主义认为自然法则支配整个宇宙）是颇具开创性的。[5]

> "有些斯宾诺莎的仰慕者可能会觉得班内特的这种质询过于无礼了。确实，斯宾诺莎的演绎再严谨，班内特的评价也不高……如果我们不承认班内特解读斯宾诺莎的方式恰恰是斯宾诺莎希望自己被解读的方式的话，那才是对斯宾诺莎的不敬。对任何否认班内特提出这种逻辑问题的恰当性的人，斯宾诺莎一点耐心也不会有。"
>
> —— 埃德温·克利：《论班内特眼中的斯宾诺莎：目的论问题》

互动

斯宾诺莎性格温和，然而一生中曾多次处于争议的中心。其中最著名的就是他在24岁时被阿姆斯特丹的马拉诺犹太人社区革出教门，理由是他思想激进，拒绝妥协；在他看来，哲学家的职责就是做出理性的解释。因此，虽然他写作《伦理学》不是为了专门去进行反击，但是对许多权威的思想观点来说却是一种彻底的、直接的、蓄意的挑战。《伦理学》质疑目的论思维（在一切事物的原因中找到一种目的）、自由意志的观念以及上帝与世界相分离的观点，这在当时是激进的，时至今日依然如此。确实，斯宾诺莎的民主诉求挑战了秩序和等级制这些传统观念的核心。斯宾诺莎也被看作环保主义非常早期的支持者。[6]

《伦理学》以一种与其他学科（例如心理学和政治学）相联系的方法来研究哲学，但是它所采用的各种方法中，除了几何学方法，其他都不能简化为某种特定风格。也就是说，形式对这部作品的内容至关重要，也是《伦理学》能够无可比拟、永远流传的部分原因，尽管有时对读者来说这种形式有些费解。

如乔纳森·班内特这样的研究者所指出的，《伦理学》最重要的遗产之一就是：这本书的失败之处也颇具启发性，至少会迫使其他人给出更好的方法来解决这些失败之处。[7]

持续争议

《伦理学》对于后世的贡献主要在于它创造性地证明了支配自然和上帝的规律是永恒的。这种观点错综复杂、应用面广，因此给一系列课题带来了灵感。当然，这些课题也面对着不少批评。但就《伦理学》本身来说，今天已经没有什么思想家的首要目标是批评斯宾诺莎了。这并不是说斯宾诺莎的观点现在已经无可指摘了，更不是说《伦理学》已经毫无吸引人之处。例如，宗教哲学家马蒂恩·布伊斯最近在他的文章《如何创造一个活生生的神》中表示，斯宾诺莎论述自由时的缺陷在于其拒绝承认运气的因素。在这一方面，斯宾诺莎经常被拿来和德国唯心主义哲学家弗里德里希·谢林进行不公正的比较，[8] 后者对形而上学态度更为公开。谢林于 19 世纪上半叶开始写作，比斯宾诺莎晚了一个半世纪。重点在于，斯宾诺莎的思想已经被现代哲学界广泛而深入地吸收了。因此，上述对他的作品的当代解读大部分都是在利用《伦理学》的创新之处，而不是在整体上挑战它。换句话说，这本书太过独特，以至于很少有人直接批评它。

如果有人要批评《伦理学》，可能的动机无非以下几条。最直接的是针对这本书的神学立场。不管是支持"上帝和世界相分离"的人，还是赞同"宇宙有其目的"的人，都不会同意这种神学立场，因为这本书抨击了这两种观点。[9] 另一类批评以实用主义 * 方法为基础。这种方法通过观察哲学论断的实际应用结果来判断它们是否正确。斯宾诺莎的体系看似和实用主义是冲突的，因为他的方法完全是自我指涉的；不管是论及自然还是人类，这种方法中永恒的法则都不允许有缺漏或者改动的可能。

---

1. 本笃十六世：《真理中的博爱》，第 34 节，安吉拉·C. 米切利："非传统的基础：与本笃十六世教皇及现代性的对话"，《政治学争鸣》第 41 卷，2012 年第 1 期，第 27 页。

2. 约书亚·J. 麦克尔维："教皇方济各：我更爱朴素的教堂"，《国家天主教报道》，2013 年 3 月 16 日，登录日期 2013 年 5 月 30 日，http://ncronline.org/blogs/pope-francis-i-would-love-church-poor。

3. 麦克尔维："教皇方济各"。

4. 参见米切利："非传统的基础"。

5. 莱奥纳多·博夫："本笃十六世正在将教会领入歧途",《国际新闻》, 2007 年 9 月 13 日, 登录日期 2013 年 9 月 26 日, http://www.ipsnews.net/2007/09/pope-benedict-xvi-is-leading-the-church-astray/。

6. 西蒙·C. 金："作为维吉尼奥·伊利佐多和古斯塔沃·古铁雷斯宗教学方法的语境神学", 博士学位论文, 美国天主教大学, 2011 年, 第 300—308 页。

7. 参见保罗·E. 西格蒙德:《十字路口的解放神学: 民主或革命? 》, 组约: 牛津大学出版社, 1980 年; 弗雷德里克·桑塔格:"政治暴力与解放神学",《福音神学学会杂志》第 33 卷, 1990 年 3 月第 1 期, 第 85—94 页。

8. 约瑟夫·A. 瓦拉卡利："古斯塔沃·古铁雷斯《解放神学》的天主教社会学批判",《天主教社会学评论》, 1996 年第 1 期, 第 175 页。

9. 金:"语境神学", 第 295 页。

12 未来展望

要点 🔑

- 《伦理学》依然备受关注，人们尤其关注其在两个领域里的应用：科学与宗教自然主义、社会与政治理论。

- 《伦理学》提出的种种观点被应用于各种各样的领域，包括政治理论、心理分析、认知科学和动物行为学（从生物学的角度研究人类行为和社会行为）。

- 因为《伦理学》对泛神论、一元论和决定论的探索，也因为它对"如何过上美好生活"这一问题的独特回答，这本书成为哲学史上的关键著作。

潜力

本尼迪克特·德·斯宾诺莎的《伦理学》仍然在吸引着强烈的关注，人们对它的兴趣方兴未艾。但兴趣如何转化为影响力就是另外一个问题了，这既取决于作品中被分析的部分，也取决于某个评论所指向的课题与目标。从这里可以延伸到哪个方向呢？有两个引人注目的领域特别突出。第一个领域是科学与宗教自然主义，第二个是社会与政治理论。例如，斯宾诺莎的自然主义就被一些当代科学哲学家作为理论资源。[1] 斯宾诺莎拒绝把上帝视为一个对世界起作用的外部能动者，这些科学哲学家中的一些人很赞赏这种看法。同时，宗教哲学家们，特别是罗伯特·科灵顿，认为斯宾诺莎的自然主义使得自然接受神学质询。[2] 政治学方面，《伦理学》形式上和内容上的反等级制基调，连同对各种目的论思想的抵制，都帮助许多当代法国思想家对现存的关于社会制度和语言的思维方式进行激烈批判。[3]

> "虽然斯宾诺莎的观点可能有些晦涩，但毫无疑问他一直在尽心尽力地阐释我们的日常经验。斯宾诺莎的形而上学和认识论为一种人类学开辟了空间，它包括人类本质的哲学和人与人之间如何产生联系的理论。斯宾诺莎赋予了我们理解自己的工具和幸福生活的策略。很少有古希腊以后的哲学家能做到这些。"
>
> —— 蓓丝·罗德："导论"，《哲学之外的斯宾诺莎》

未来方向

《伦理学》被人们接受的历史充满了各种波折。最初，人们因为它的无神论观点而抵制它，在泛神论论争之后，它又被作为关键的哲学著作获得重新评价。《伦理学》的见解涉及诸多领域，从形而上学到认识论，再到政治理论、心理分析、动物行为学，不一而足。斯宾诺莎的影响形式多变，难以预测。从这一点来说，当我们讨论《伦理学》研究的未来方向时，必须要保持思想开放。

斯宾诺莎思想最近一个出乎意料的应用出现在神经学家安东尼奥·达马西奥[*]的研究中。与认知科学（对心智的研究）的主流不同，他的研究为情状（即情感）找到了神经科学的解释。在有关严谨科学研究的一系列科普书中，达马西奥否定了近来神经科学研究中对认知（思维）进行强调的笛卡尔式方法，[4]相反，他推进了对情状研究的斯宾诺莎式方法。[5]达马西奥直接借鉴斯宾诺莎的心理学原理，创立了快乐和悲伤的神经学（就是研究这些情感的神经学基础）。虽然不同立场的研究者（包括哲学家）已经对达马西奥的独特观点有所批评和否定，[6] 但是对情状进行神经科学分析这项工程将会持续繁荣。这项工程与技术的进步有关，例如情状计算（研究通过电脑来模拟情状）。达马西奥对斯宾诺莎的理解可能并不全部正确，[7] 但他确实在这些问题的讨论中表达了自己的见解，迈出了一大步。

小结

只要哲学家们仍在争论心灵、存在、知识或伦理学等话题，本尼迪克特·德·斯宾诺莎就会有读者。斯宾诺莎对于哲学中长期存在的问题有着独特的见解，这影响了足够多的和他一样有名望的哲学家，也保证了《伦理学》会出现在未来的哲学著作中。当代哲学的主要思潮都要求仔细阅读《伦理学》。这说明，即便是这本书里面的某些复杂细节都有可能成为现在研究的主题。

斯宾诺莎的生平也值得我们持续关注。在一个对政治和宗教持非正统的观点就会招来牢狱之灾乃至杀身之祸的时代，他也从不妥协，将追求真理、探究自然和使人类获得幸福视为己任。年轻的时候遭到马拉诺社区驱逐之后，他宁可在荷兰乡村以磨镜片为生，过着与世隔绝的生活，也不肯放弃自己激进的观点。许多与之有私交的人，即便和他意见相左，都证实他是一个温和亲切的人。

斯宾诺莎的《伦理学》风格独特，涉及内容广泛，观点激进，受到了古典、中世纪和当时思想界的不同影响，斯宾诺莎将这些影响共冶一炉，凭借自己的方法论形成了完整的哲学体系。在这个体系中，斯宾诺莎提出了一个问题：我们如何在一个没有自由意志的、人类不居首位的、由决定论支配的世界里获得最大的幸福？《伦理学》中的观点和这些观点被呈现的形式是分不开的，那是一种由定义和命题组成的复杂的和自我指涉的框架。斯宾诺莎的目标就是从新的角度解读心灵、身体、自然和上帝，并把它们化为单一的实体，这成了西方传统中最激进的研究方法之一。斯宾诺莎的上帝不是有情感、有目标、有计划、能够以自由意志在世界上行动的存在，它不具备这些属性。他的上帝是一个单一实体，把身体和精神领域合二为一。这个实体也是永恒不变的法则，它中立地规范着包括人类在内的万事万物。

1. 参见马乔里·G.格林和黛博拉·内尔斯编：《斯宾诺莎与科学》，多德雷赫特：克鲁沃出版社，1986年。

2. 罗伯特·S.科林顿：《狂热的自然主义者：世界的迹象》，印第安纳波利斯：印第安纳大学出版社，1994年。

3. 参见路易·阿尔都塞：《保卫马克思》，本·布鲁斯特译，伦敦：沃索出版社，1969年，第78页；吉尔·德勒兹：《斯宾诺莎的实践哲学》，罗伯特·赫利译，旧金山：城市之光出版社，1988年，第69—70页。

4. 安东尼奥·R.达马西奥：《笛卡尔的错误：情绪、推理和人脑》，纽约：哈珀出版社，1995年。

5. 安东尼奥·R.达马西奥：《寻找斯宾诺莎：快乐、悲伤和感受着的大脑》，佛罗里达州奥兰多：哈威斯特出版社，2003年。

6. 参见M.R.贝内特和P.M.S.哈克：《神经科学的哲学基础》，马萨诸塞州马尔登：威利-布莱克威尔出版社，2003年。

7. 伊恩·哈金："小心大脑"，《纽约书评》，2004年6月24日，http://www.nybooks.com/articles/archives/2004/jun/24/minding-the-brain/。

术语表

1. **利他主义**：为他人的幸福着想。

2. **神人同形同性论**：把人的特性套用到非人类的事物上（例如，把车灯想象成车的眼睛，或者把神想象为巨人）。

3. **无神论**：一种否认上帝存在的信念。

4. **定理**：一种在其基础上可以得出其他推论的、不证自明的假设。

5. **笛卡尔的**：源自勒内·笛卡尔，用来指称受笛卡尔作品影响的观点。

6. **因果关系**：在哲学中指对原因和结果间关系的研究。

7. **当灭之物**：希伯来语术语，指被革出教门、被禁绝来往、被所在的信仰团体驱逐。

8. **欲求**：斯宾诺莎作品中的术语，指的是每一个事物努力生存的那股力量。

9. **决定论**：一种哲学观点，认为所有事物都由无法产生任何其他行动的原因决定。决定论否定自由意志的存在，因为决定论者认为人的行动被决定的方式和其他的行动一样。

10. **经验主义**：哲学的一个分支，强调感官经验，比如看或听，认为这才是知识的首要或者是唯一的来源，也指通过观察和实验获得的知识。

11. **启蒙运动**：一场思想界的运动，主要发生在 18 世纪。它捍卫理性，挑战传统的政治权威，呼吁在大众中传播知识。斯宾诺莎经常被视为激进启蒙运动的早期推动者。

12. **认识论**：哲学中的一个领域，处理与自然和知识的界限相关的各种问题。

13. **动物行为学**：从生物学的角度研究人类行为和社会组织的科学。

14. **革出教门**：遭所在的信仰团体禁绝来往或驱逐。

15. **目的因**：亚里士多德讨论过的一种原因。除了其他种类的原因之外，他还区分了效力因和目的因。效力因这个概念就是现在普遍意义上的原因：某种程度上如果 A 发生在 B 之前，A 是 B 的原因，那么 B 就是它的产物或是结果。但目的因（希腊语 *teloi*）指的是一件事物为的是什么，即它的目的。例如，一艘船的效力因可能是造这艘船的船匠，但它的目的因可能是在海上航行。

16. **几何学方法**：欧几里得几何学中首创的一种特有的证明方法，被斯宾诺莎应用在哲学写作上。这种方法一般先给出定义和定理，再开展论证和证明。

17. **德国唯心主义**：指的是随着伊曼努尔·康德《纯粹理性批判》的问世而产生的一种德国哲学传统。广义来说，从属于这一传统的思想家，如格奥尔格·威廉·弗里德里希·黑格尔和弗里德里希·谢林，努力探究的是心灵与这个世界互动的条件。

18. **异端**：与某一种宗教中公认的神学观点相偏离。

19. **马拉诺人**：伊比利亚半岛（西班牙和葡萄牙）的犹太人后裔，在 15 世纪当地穆斯林摩尔人被击败之后，曾被命令改信基督教。斯宾诺莎的家族是葡萄牙马拉诺人，之前被称作"埃斯皮诺萨"，16 世纪 90 年代移民到阿姆斯特丹。

20. **门诺派**：基督新教的一个教派，以门诺·西门命名，起源于 16 世纪。他们拒绝给婴儿洗礼，而实行对成年信众的洗礼，这招致了来自其他基督教团体如天主教和新教的蔑视和迫害。门诺派以他们的非战主义信仰闻名。

21. **形而上学**：源自古希腊哲学，特别是源自亚里士多德的哲学思想。这是一个探究存在本质的哲学分支。

22. **一元论**：一种基本上认为只存在一个实体或事物的哲学立场。实体一元论认为只有一个实体存在。

23. **自然主义**：一系列哲学观点，其共同的基本理念是，自然法则是解释一切现象的基础。

24. **能动的自然和被动的自然**：译自拉丁语的术语，分别意为"主动意义上的自然"和"已经被创造出来的自然"，更准确的译法是"能动的自然"和"被动的自然"，暗指因（能动的自然）与果（被动的自然）或创造者与创造物之间的关系。

25. **本体论**：哲学中处理与存在有关的问题的部分，即：从广义上来说，有或者存在什么样的事物？或：现实是由什么构成的？

26. **万有在神论**：认为"上帝在存在着的万物里"的一种观点。它与泛神论不同的是，后者认为上帝就是宇宙中的万物（上帝是自然），而万有在神论认为上帝在宇宙中的万物中（上帝在所有自然中）。

27. **泛神论**：认为"宇宙中的万物等同于神"的观点。虽然这个术语本身并没有出现在《伦理学》里，但这本书被认为是最重要的泛神论哲学文本之一。

28. **泛神论论争**：18 世纪 80 年代戈特霍尔德·莱辛和弗里德里希·亨利希·雅科比之间的一场论争。莱辛赞同斯宾诺莎的哲学，把它当作不同于当时权威思维模式的一种选择。这场论争所提出的问题不仅改变了人们对斯宾诺莎的看法，同时也标志着德国思想界的一次关键转变。德国思想界从 18 世纪的启蒙运动时期转向 19 世纪的浪漫主义和唯心主义时期。

29. **实用主义**：19 世纪 70 年代发源于美国的一种哲学传统。关键人物包括查尔斯·桑德斯·皮尔士、威廉·詹姆斯以及约翰·杜威。这些人都致力于在生活经验里观察和阐明哲学论断的作用。

30. **理性主义**：启蒙主义运动思想的一个分支，认为理性既是获取知识的手段又是真理的贮藏库。理性主义在 17 世纪晚期和 18 世纪达到了其影响力的高峰，与认为知识可以通过观察和试验获得的经验主义形成对立。

31. **宗教改革**：1517 年马丁·路德大张旗鼓地把《九十五条论纲》贴在了维登堡大教堂的大门上，宗教改革运动就此开始了。这最终导致了天主教与新教的分裂。

32. **文艺复兴**：14 世纪到 17 世纪欧洲历史上的一段时期，连接了中世纪和现代历史。它开始于意大利，是一场复兴古希腊文化的思想、

文艺运动。Renaissance 的字面意思是"重生"。

33. **思想物和广延物**：拉丁语中的笛卡尔术语"思想的事物"（思想物）和"被延伸的事物"（广延物），用来说明心灵的重要属性是以思想为特征的，身体的重要属性是以空间中的延伸为特征的。

34. **浪漫主义**：18 世纪末 19 世纪初在欧洲兴起的哲学、艺术和文学运动，反抗理性主义看重的普适性、理性和数学运算，强调历史的特殊性、情感以及自然世界的精神特性。

35. **经院哲学**：一种在中世纪欧洲大学内部占主导地位的哲学方法，其特征是逻辑辩论，并以此为方法进行尽可能微妙的区别。

36. **评注**：希腊语 *scholia*，在几何学的写作风格里，斯宾诺莎会用这种形式来写一些非正式的评论。

37. **智力直觉**：这个术语可以译作"智性直观"，按照斯宾诺莎的观点，它是通过感觉或纯粹理性获得的知识以外的第三种知识。

38. **斯多葛派**：由基提翁的芝诺创立的希腊哲学学派，他们认为一个人能控制自己的激情是人生中的至善。

39. **宗教议会**：教会的议会或大会，通常由神职人员中的高级成员组成。

40. **目的论**：起源于希腊语 *telos*，意思是"目标"或"目的"，既指对目的的研究，又指任何一套认为"目的是事物或事件的原因的一部分"的论断。

人名表

1. 亚里士多德（公元前384—322年），希腊哲学家，雅典学园创建者，柏拉图的学生。在几乎每一种人类知识的领域都撰写过专著，包括伦理学、美学、形而上学和逻辑学。

2. 阿尔伯特·巴尔兹（1887—1957），美国当代哲学家、早期现代哲学史研究者。

3. 皮埃尔·培尔（1647—1706），法国哲学家。继笛卡尔之后，他主张将信仰和理性严格分开。他的《历史批判辞典》于1697年首次出版，在17世纪和18世纪之交时的欧洲知识分子中有大量的读者。

4. 乔纳森·班内特（1930年出生），英国当代哲学家、早期现代哲学史研究者。

5. 豪尔赫·路易斯·博尔赫斯（1899—1986），阿根廷作家、诗人，20世纪文学领域中最重要的人物之一。

6. 罗伯特·S.科灵顿（1950年出生），美国当代作家、哲学家。

7. 埃德温·克利（1937年出生），美国当代哲学家、斯宾诺莎研究学者。

8. 安东尼奥·达马西奥（1944年出生），当代神经学家、作家。

9. 吉尔·德勒兹（1925—1995），法国当代哲学家、哲学史家。

10. 勒内·笛卡尔（1596—1650），法国数学家和哲学家。他提出心灵与世界的二元论，采用哲学方法在不容置疑的基础问题上得出清晰、确切的观点，这些都是中世纪哲学到现代哲学的决定性转折点，也造成了存在哲学到意识哲学的普遍转向。

11. 阿尔伯特·爱因斯坦（1879—1955），20世纪理论物理学最重要的贡献者，创立了相对论。

12. 弗朗西斯科斯·范·登·恩登（1602—1674），政治激进主义者、诗

人、哲学家，因曾是斯宾诺莎的老师而闻名。他曾经做过一段时间耶稣会信徒，但他很快就放弃了信仰，因此有人认为他是一个无神论者。他因密谋刺杀法国国王路易十四而被定罪，绞死在巴士底狱外。

13. 亚历山大里亚的欧几里得（约公元前 4 世纪到公元前 3 世纪中期），希腊数学家，被公认为几何学之父，最重要的贡献是《几何原本》一书。直到 20 世纪早期这本书依然是数学课的教材。

14. 西格蒙德·弗洛伊德（1856—1939），奥地利神经学家，心理分析学之父。在思想史上，弗洛伊德和卡尔·马克思、弗里德里希·尼采一样重要。他们瓦解了启蒙运动建立的"人从根本上来说是理性的存在"这一共识。

15. 彼得·盖伊（1923—2015），美国当代历史学家。

16. 约翰·沃尔夫冈·冯·歌德（1749—1832），德国哲学家、作家、科学家和政治家。他是 18 世纪晚期"狂飙突进"文学运动的代表人物。

17. 菲利克斯·伽塔利（1930—1992），法国当代心理治疗师、哲学家和活动家。

18. 格奥尔格·威廉·弗里德里希·黑格尔（1770—1831），过去两百年间最重要的哲学家之一。黑格尔派的唯心主义叫做绝对唯心主义，与伊曼努尔·康德创立的德国唯心主义传统有关。绝对唯心主义把自然描绘为一种绝对的体系，它的表现跨越了逻辑历史的路径。

19. 弗里德里希·亨利希·雅科比（1743—1819），后康德时代的德国哲学家，曾参与泛神论论争。

20. 戈特弗里德·威廉·冯·莱布尼茨（1646—1716），德国数学家、哲学家。他和艾萨克·牛顿都被认为是微积分的独立的共同发明人。

21. 戈特霍尔德·莱辛（1729—1781），德国哲学家、诗人。在德国启蒙主义思想与浪漫主义的过渡时期，他的作品被认为是一个巅峰。

22. 约翰·洛克（1632—1704），英国哲学家、政治理论家。他的作品被认为是经验主义哲学和现代自由主义的基础。

23. 迈蒙尼德（1135—1204），本名摩西·本·迈蒙，是 12 世纪哲学家、《律法书》学者，居住在西班牙和北非。他因阐释犹太律法、将亚里士多德思想和圣经思想结合起来而声名远播。

24. 尼古拉斯·马勒伯朗士（1638—1715），笛卡尔派的理性主义哲学家、牧师。他最有名的是对某一种偶因论的辩护。偶因论认为精神事件不会导致身体事件，反之亦然；而每次心灵有什么意图，是上帝让身体去实行的。

25. 皮埃尔-弗朗索瓦·莫罗（生于 1948 年），法国当代哲学家、法国哲学史研究者。

26. 弗里德里希·尼采（1844—1900），德国哲学家、文化批评家。他的作品彻底挑战了许多西方思想中最为根深蒂固的概念，特别是那些与道德和宗教有关的概念。

27. 伯特兰·罗素（1872—1970），英国当代哲学家、数学家，也是 20 世纪最重要的公共知识分子之一。

28. 弗里德里希·谢林（1775—1854），德国哲学家，与伊曼努尔·康德（1724—1804）创立的德国唯心主义传统有关。他强调运气是自然中不可逃避的事实。

29. 扬·德·维特（1625—1672），共和主义政治家，从 17 世纪 50 年代至逝世担任荷兰共和国的第一任国家元首。1672 年他被保皇党废黜，和他的哥哥一起被处以私刑。

WAYS IN TO THE TEXT

- Baruch Spinoza (1632–77) was a Dutch philosopher. His *Ethics* was so radical that he was excommunicated* from his Jewish community, but the book is today considered a key work of modern philosophy.

- *Ethics* argues that everything is ultimately an expression of a single substance, which Spinoza calls God or, what is the same, nature. He rejects the common idea of an anthropomorphic* God (one who possesses human qualities, and who can love or punish people).

- *Ethics* shows that it is possible to think about how to live well in a world that has no purpose or pride of place for human beings.

Who Was Spinoza?

Baruch Spinoza was born in Amsterdam in 1632. His family were Marranos,* Jews who had fled Portugal after being persecuted. He received his education in the Marrano community of Amsterdam. The same community excommunicated him in 1656, apparently due to his controversial theological views. He went on to live in relative isolation, working as a lens-grinder in the small town of Rijnsburg and later in Voorburg and the Hague.

Following his excommunication, Spinoza took the Latinized form, Benedictus, of his Jewish name (Baruch), and wrote most of his philosophical texts using the new name. A particularly notable exception to this is the most controversial work published during his lifetime, *Theologico-Political Treatise*, an anonymous publication from 1670. The Synod,* or governing council, of the

Dutch Reformed Church banned the work in 1674, by which time it was clear that Spinoza was its author.

Published after Spinoza's death, *Ethics* was almost universally condemned upon its publication. For a long time, philosophers would try to avoid association with Spinoza's radical ideas. It wasn't until a century or so after his death that his work began to be seen in a positive light in the context of philosophical debates.

Spinoza's bad reputation was, to some extent, the result of his own steady devotion to principles that went against the grain of the times in which he lived. Some of the things he fought for during his lifetime are commonly taken for granted in today's Western world, however, including his defense of religious tolerance and basic civil liberties.

What Does *Ethics* Say?

The emergence of modern science in the late Renaissance* led to the discovery that the earth is not at the center of the universe. This called for a radical reevaluation of our understanding of humanity's place in the cosmos. Spinoza's *Ethics* offers a guide to living in a universe in which human beings are bound by the deterministic* laws of nature—in other words, just as the natural world is governed by strict laws of science, the human world is regulated by laws that people cannot overturn through their supposed free will. Spinoza passes through various stages in order to finally reach his answer to the question of how to live well in a deterministic universe. The overall approach of *Ethics* is based on the geometrical method* first developed by the ancient Greek mathematician Euclid:* it begins with a series of definitions of Spinoza's terms coupled with a series of axioms* (self-evident claims), from which Spinoza seeks to demonstrate a number of proofs. His proofs cover a range of topics in philosophy and other disciplines, beginning with metaphysics* (the study of the nature of being) and theology, and moving on to epistemology* (the investigation of the nature and limits of knowledge), psychology, the theory of action (Spinoza's analysis of what it means to act), and finally an account of human freedom.

In his metaphysics, Spinoza develops a form of pantheist* monism,* according to which God is identical with nature. (Monism states that all existence is an expression of one single underlying entity; pantheism identifies that one entity as God.) He argues that the whole of reality is a manifestation of God, which he sees as the one entity underlying all of reality, and he uses the technical term "substance" to refer to this entity. Spinoza's God (or nature) is this one substance, which is not shaped in the image of man. God is not anthropomorphic (does not possess human qualities), and therefore God is not, as various religions hold, a being who is capable of loving humanity, or able to punish humans for their sins.

Extending his monism, Spinoza denies the existence of free will. He argues that human freedom involves a recognition that the world functions in a way that is deterministic (that is, every event follows from a cause that makes it necessary and inevitable). Based on this, *Ethics* offers an intriguing analysis of human emotions, which are seen as stemming from joy, sadness, and desire. These, in turn, he views as expressions of the effort to persist in being (the drive to survive and prosper). Spinoza argues that the blessed life lies in an intellectual love of God by people, despite the fact that God cannot love human beings in return. This intellectual love for God involves a knowledge and understanding of deterministic causation.*

Spinoza's thought was too radical for his contemporaries. His denial of free will and rejection of the anthropomorphic view of God were not ideas that seventeenth-century thinkers were ready to accept. Spinoza's thought became influential only after a debate over pantheism in late eighteenth-century German philosophy. This would later lead G. W. F. Hegel* to claim that, "You are either a Spinozist or not a philosopher at all."[1]

Why Does *Ethics* Matter?

Spinoza wrote *Ethics* in a manner that is often difficult to understand. Despite this, philosophy students who wish to follow the development of modern debates in metaphysics and epistemology are bound to read it. It contains, for instance, a significant and highly original discussion of the relationship between mind and body—which he sees as two distinct but parallel expressions of the same underlying substance. According to Spinoza's view, each event has parallel bodily and mental expressions.

Spinoza's work also contains some intriguing insights into the psychology of what we ordinarily think of as emotions, which he referred to as "affects." It offers a definition and in-depth discussion of love, hate, hope, fear, and other affects. Spinoza considers all these to derive from the three basic affects of joy, sadness, and desire. These are states of the body reflected in the mind. Furthermore, they are states that can take over the mind, preventing rational thought. Spinoza argues that having knowledge of the mechanisms at work behind such emotional states allows us to control them. This, in turn, is the foundation of living the good life. This view reflects the influence of the ancient Stoics* on Spinoza. The Stoics also considered the regulation of emotion as a pathway to the good life.

The above highlights the various levels of analysis at work in Spinoza's text. It begins with an abstract discussion of metaphysics and ends with a particular discussion of ethics. Between these comes philosophy of mind, epistemology, psychology, a theory of action, and a discussion of freedom. Various aspects of Spinoza's thinking have influenced people working in many different fields. From these lines of influence, many new strands of thinking may develop.

---

1. Merold Westphal, "Hegel between Spinoza and Derrida," in *Hegel's History of Philosophy: New Interpretations*, ed. David Duquette (Albany: State of New York Press, 2003), 144.

SECTION 1
INFLUENCES

THE AUTHOR AND THE HISTORICAL CONTEXT

KEY POINTS

* Spinoza's book is a key text of modern philosophy. It addresses the question of how to live a good life in a deterministic* universe—one in which human action is not guided by free will.

* Spinoza was expelled from the Jewish community of Amsterdam for holding views that went against their established beliefs. He moved elsewhere in Holland and produced great works.

* Spinoza supported Dutch liberalism of the seventeenth century, which came to an abrupt end in 1672 when the liberal government was overthrown by royalist forces.

Why Read This Text?

Benedictus de Spinoza's *Ethics* is one of the most significant books in the history of philosophy. Written in Latin between 1661/2 and 1675 and published in 1677 following its author's death, the work owes its title to the fact that it addresses questions of right and wrong human action. Yet this title is misleading, as *Ethics* is also a book on God and the world, an examination of the structure of the mind and its relation to the body, a detailed account of the psychology of emotions, and an attempt to understand human action. *Ethics'* rejection of free will and exploration of the consequences of determinism—the philosophical view that all human action is determined by causes outside of free will—remains radical to this day. The book argues that altruism*—concern for the welfare of others—is a consequence of, rather than being contradictory to, the pursuit of self-interest. Its insightful analysis of human emotions concludes with an account of the good life as one characterized by joy and the power for action. The work is demanding for the reader, but is clearly a key text of modern philosophy.

The questions that *Ethics* addresses, including what the universe is, how we know it, how our emotions relate to it, and how we should act within it, are as important today as they were in Spinoza's time. It remains a lasting credit to *Ethics* that it addresses them in such a systematic and original way. Even the work's geometric style is not accidental. It reflects the author's belief in universal laws that govern everything equally and without preference or gaps.

> *"By the decree of the angels, and by the command of the holy men, we excommunicate, expel, curse and damn Baruch de Espinoza ... the Lord will blot out his name from under heaven ... But you who cleave unto the Lord God are all alive this day. We order that no one should communicate with him orally or in writing, or show him any favor, or stay with him under the same roof, or within four ells of him, or read anything composed or written by him."*
>
> —— "Excommunication of Spinoza by the Talmud Torah congregation of Amsterdam," quoted in Steven Nadler, *Spinoza: A Life*

Author's Life

Spinoza was born in Amsterdam, Holland, in 1632 and given

the Hebrew name Baruch. Raised among that city's Marrano* community, he worked for a period in his family's importing business, while at the same time showing promise as a scholar. Little is known about Spinoza during those early years, which seem to have involved a degree of controversy—most likely due to his controversial theological views. Such unorthodoxy was probably the reason why he was subjected to a knife attack in 1656.[1] On July 27 of that same year he was excommunicated* from his community in a writ of *cherem*,* the Hebrew term for expulsion of a member of the community. This followed accusations against him of "horrible heresies" and "monstrous actions."[2] The reasons remain unclear, but are likely to stem from his unorthodox views on prophecy, immortality, and God. Those who had lived, worked, and studied alongside him were then forbidden to associate with him. In 1661, Spinoza left Amsterdam for the village of Rijnsburg, near Leiden, in Holland, where he began work on *Ethics*. At the same time, he Latinized his name to Benedictus. In Rijnsburg, as well as across later moves to Voorburg and the Hague, Spinoza supported himself in his philosophical work by grinding lenses for optical instruments.

In exile, Spinoza corresponded with a network of friends and supporters who built his reputation and circulated early drafts of *Ethics*, at least across Holland and England. This group, many of whom were Mennonites,* members of non-conformist Protestant sects, attracted the suspicion of political and Church authorities. Those authorities grew increasingly hostile toward certain intellectual circles—if not wider culture—during the seventeenth century in Europe, particularly in England and the Netherlands. His friends encouraged him to publish *Ethics*, but Spinoza hesitated, discouraged by his excommunication and the hostile reaction to another work, his *Theological-Political Treatise*, published anonymously in 1670 and containing many similar ideas to those in *Ethics*.

Spinoza died from tuberculosis in February 1677. His death was possibly due to inhalation of glass, an effect of his work grinding lenses. *Ethics* was published later in the same year.

Author's Background

Spinoza's works were written and published during a turbulent period in the history of Holland. Political liberalism and religious tolerance played a central role in the Dutch Republic, which was headed by Jan de Witt* at the time Spinoza was writing. Spinoza was in fact granted a small pension by de Witt during the height of the latter's political career in the 1670s.[3] It was the general climate in the Netherlands, seemingly favorable to the circulation of liberal ideas, that made it possible to publish a number of controversial ideas put forward in Spinoza's works. Many of these are to be found in his *Theologico-Political Treatise* (*Tractatus Theologico-Politicus*, first published in 1670), which can be read as a defense of de Witt's political project. It should be noted, however, that even in this tolerant climate, Spinoza seems to have felt uneasy about publishing such controversial views in his own name.

Spinoza's caution would soon prove justified, as the fate of his work ended up being tied to that of de Witt's leadership. In 1672 (known in Dutch as the *Rampjaar* or Disaster-year), as a result of the Franco-Dutch War, de Witt and his followers were overthrown and lynched by royalists. Spinoza's anonymity was lost, and in this climate his work soon came under attack as being "forged in Hell by a renegade Jew and the Devil, and issued with the knowledge of Jan de Witt."[4] This led to it being condemned and banned by the Dutch Reformed Church in 1674.[5] *Ethics*, published after Spinoza's death, would come to be attacked in a similar way during the following century. This was a time when being described as a follower of Spinoza would amount to a kind of insult among philosophers. It was only around the late eighteenth century, and not without a measure of hesitation, that it became permissible to include Spinoza's work as part of serious philosophical debate.

---

1. First reported by Pierre Bayle in his biography of Spinoza; see: H. M. Ravven and L. E. Goodman, *Jewish Themes in Spinoza's Philosophy* (New York: SUNY Press, 2012), 269.

2. Genevieve Lloyd, *Routledge Philosophy GuideBook to Spinoza and the* Ethics (London: Routledge, 1996), 1.

3. See: Roger Scruton, *Spinoza* (Oxford: Oxford University Press, 1986), 11.

4. Scruton, *Spinoza*, 11.

5. Scruton, *Spinoza*, 11.

MODULE 2
ACADEMIC CONTEXT

KEY POINTS

* Spinoza's thinking reflected the events of his time. Europe's religious wars led him to value the idea of tolerance, and the growing scientific revolution led him to question religious explanations about the world.

* René Descartes,* a leading influence on Spinoza, proposed that the mind and the body exist independently. Yet his theory of how they interact was considered inadequate by later philosophers.

* Spinoza rejected Aristotle's* notion of final cause. He also differed from Descartes in proposing that mind and body are two aspects of a single underlying substance (rather than two distinct substances).

The Work in Its Context

Benedictus de Spinoza's *Ethics* was written during a period of change in European thought. Not long before, Europe had undergone the violence of the continent's post-Reformation* religious wars, and ideas of religious tolerance were forming in places such as the Netherlands, where Spinoza lived. The discoveries of modern science were, by now, in the process of replacing the old medieval scholastic* establishment associated with the Roman Catholic Church. The demise of scholasticism came with the collapse of a number of supposed certainties (the belief, for instance, that the earth is at the center of the universe).

Spinoza was affected both by the political developments of

his day and by the questioning spirit of modern science. Having personally experienced religious persecution, he had come to see the priceless value of religious tolerance, for which he would argue throughout his life. Spinoza's view of religion was informed by the new types of questioning made possible by modern science, and by the discovery that the natural world can be explained through deterministic* laws of nature—laws that regulate which effects necessarily follow from which causes. A basic feature of Spinoza's thought is the embracing of determinism, accompanied by an attempt to examine how one may live a good life in a deterministic universe.

> "At last I have discovered it—thought; this alone is inseparable from me. I am, I exist—that is certain ... I am, then, in the strict sense only a thing that thinks; that is, I am a mind, or intelligence, or intellect, or reason—words whose meaning I have been ignorant of until now. But for all that I am a thing which is real and which truly exists. But what kind of a thing? As I have just said—a thinking thing."
> —— René Descartes, *Meditations on First Philosophy*

Overview of the Field

Spinoza and his contemporaries were writing at the start of a move away from the older Aristotelian philosophical tradition— a shift led by the rise of modern science. One of the movement's leading figures was René Descartes, who helped develop modern philosophy. Descartes argued for the separation of mind from matter. He called the material world *res extensa*, meaning "the extended thing," because he believed the property of spatial extension to be the main feature of matter. *Res extensa** exists separately from the realm of the mind (*res cogitans,** the thinking thing). Mind and body are, according to Descartes, two distinct substances, and one can exist independently of the other. Descartes's argument allowed him to go on to claim, in his physics, that material objects can be explained on the basis of mechanistic laws. Material bodies are, according to Descartes, like cogs in a machine: the movement of one part of the machine inevitably causes the movement of another. The material world is governed by deterministic laws that can be discovered by physics.

The implication of Descartes's separation of mind and body is that whereas bodies can be accounted for by the laws of physics, minds are not governed by such deterministic laws. Thus minds, in contrast to bodies, can have free will. Descartes's account, however, faced a very serious problem from the start: given that minds and bodies are independent substances, how can the mind interact with the body? How is it that I can move a part of my body simply because I want to (that is, simply because my mind makes the free choice to do so)?

Descartes attempted to answer this question by proposing that a part of the body, located in the brain, interacts with the mind. This is not a satisfactory answer, however, since it relies on an exception to the separation of mind and body. Modern philosophers after Descartes, including Spinoza, sought to give alternative solutions to the so-called mind/body problem.

Academic Influences

Among Spinoza's main influences was his childhood education, which was shaped by the Jewish tradition and his exposure to philosophical figures such as the twelfth-century Jewish Bible scholar Maimonides* and the ancient Greek philosopher Aristotle. Beyond his home community, Spinoza studied Latin under Franciscus van den Enden,* a former Jesuit and political radical whose home attracted a circle of humanists and freethinkers, and to whom Spinoza's political views owe much.

In terms of the content of *Ethics*, it helps to distinguish pre-modern from modern influences. The impact of Maimonides can be seen in Spinoza's insights on the philosophical basis of law. Maimonides also helped shape Spinoza's argument that humankind is not the center of creation, as well as his identification of God as both the source of understanding and the totality of that which is understood. Another major pre-modern influence on *Ethics* was classical Stoicism.* The Stoics' claims that one ought to calmly accept what one cannot control shaped much of the latter three sections of Spinoza's work. Finally, Aristotle's philosophy provides the terms used in *Ethics* to address causation* and substance, even though the conclusions reached in both are radically different. Aristotle distinguished between two different types of causation. These he called efficient (what we would today consider a common-sense idea of one thing causing another; for instance, a boat's efficient cause is its builder) and final (a thing's purpose, for example, a boat's final cause is to sail on the water).

Spinoza argued that the only true causes are efficient, denying the existence of final causes,* and thus putting forth a radical critique of teleology,* the study of the purposes of things. Moreover, while Aristotle saw the world as composed of many different substances, Spinoza argued that everything is merely an expression of a single substance: God or, what is the same thing, nature.

Among the work's modern influences, Descartes stands out. An established expert on Descartes, Spinoza employed Cartesian (that is, "from Descartes") terms in *Ethics* more than any other thinker. Yet as with Aristotle, Spinoza utilizes them to come to very different conclusions from those reached in their original context. Descartes presented God as substance in a unique sense, independent of all other beings in a way that applies to Him alone; he retained the world as a distinct realm governed by rational laws. Spinoza, on the other hand, transformed the idea of God into the concept of a unique substance, of which all other things are modes or expressions.

THE PROBLEM

KEY POINTS

* The rise of modern science made people question ideas to do with human beings' central position in the universe. Among these was the notion that people, in contrast to other beings, have free will.

* Spinoza worked at a time when modern scientists and philosophers, such as René Descartes,* were rallying against the more conservative scholasticism* of Europe's universities.

* For Spinoza, as for other rationalist* philosophers of his day, a key, unresolved issue, was how to explain the interaction between the mind and the body, which were understood to be two separate substances.

Core Question

The core question in Benedictus de Spinoza's *Ethics* is how to act in a universe that has no purpose or free will, or pride of place for humanity. The heart of Spinoza's answer to this question lies in one of the work's most original notions, that of *conatus*,* Latin for "striving," which is the effort to persist in being (that is, to survive and prosper). Spinoza claims that to live a good life, one needs to act on the *conatus*, and follow one's self-interest.

Spinoza bases his ethical ideas on the claim that all human beings are self-interested, with each striving to persevere in his or her own being. For Spinoza, this self-interest is a virtue. It is important to note that, for him, the effort *to be* involves the exercise of the intellect (rather than, for example, a brute biological urge to struggle for survival). As such, the *conatus* is closely linked with the effort to know or understand. Although this may seem like a license for pure selfishness, Spinoza intends it, conversely, to help explain how altruism is actually a product of the pursuit of self-interest. According to Spinoza, a self-interested person seeking the good life must be altruistic. In other words, there are good reasons for a self-interested person to care about the well-being of others.

Another effect of Spinoza's notion of *conatus* is that free will is an illusion, at least in the sense that one thinks of willful action as a power of the mind over the body. As he puts it, "In the mind there is no absolute, or free, will, but the mind is determined to will this or that by a cause which is also determined by another, and this again by another, and so to infinity."[1]

"As far as good and evil are concerned, they also indicate nothing positive in things, considered in themselves, nor are they anything other than modes of thinking, or notions we form because we compare things to one another. For one and the same thing can, at the same time, be good, and bad, and also indifferent. For example, music is good for one who is melancholy, bad for one who is mourning, and neither good nor bad to one who is deaf."

—— Benedictus de Spinoza, *Ethics*

The Participants

At the time when Spinoza wrote *Ethics*, the key intellectual conflict was the struggle between scholastic* and modern thinkers over the fate of the European universities. Since the Middle

Ages, intellectual life in the universities had been dominated by scholasticism, which favored lengthy debates over the logic and metaphysics* (which examines the nature of being) of texts by Aristotle* and other authoritative writers.

By the seventeenth century, greater interest in the investigation of nature, and horror at the continent's bloody experiences of religious conflict, had prompted a search for new methods through which to establish knowledge. Such attempts emphasized observation and experimentation, and styles of communication that sought to be universal—cutting across cultural differences. Along with René Descartes, who came before him, Spinoza allied himself firmly with the modern philosophers, and *Ethics* stands at the forefront of this intellectual shift. By the time Spinoza began to write, the modern Cartesian rationalist approach to philosophy had already started to be established as an alternative to scholasticism. Rationalism, in contrast to empiricism*—which held that all knowledge is acquired through observation—maintained that some knowledge is derived through reason alone. Apart from Descartes, who instigated this movement, leading representatives of rationalist philosophy included Nicolas Malebranche* and, at a later stage, Gottfried Wilhelm von Leibniz.*

The Contemporary Debate

Spinoza lived and worked at a time when the rise of modern science had led people to question Aristotle's scholasticism. One central Aristotelian concept under scrutiny was the notion of "final cause."* According to Aristotle, in order to fully understand a phenomenon under investigation, one needs to understand its purpose or final cause (in Greek, its *telos*), which is the answer to the question "what is it for?" Thus, an Aristotelian natural philosopher would consider an examination of purposes (teleology*) to be a crucial part of all science. Modern science would hold no place for teleological explanations. Descartes, for example, argued that material things can be fully explained on the basis of deterministic (or "mechanistic," as he called them) laws of nature, which require no reference to purpose for their explanation. Yet Descartes also argued that, apart from material things, there are also mental things, which are not subject to mechanistic laws.

Spinoza's metaphysics relies on his discussion of the concept of "substance," which he inherits directly from Descartes. Descartes in turn introduces it into modern philosophical discussions by radically revising earlier Aristotelian conceptions of substance. According to Descartes, a substance is a thing whose existence does not depend on the existence of any other thing. Descartes divides substances into two kinds: infinite and finite. There is only one infinite substance, according to Descartes, which goes under the name of God. God is self-caused and needs no other entity in order to exist. By contrast, a finite substance is one whose existence is dependent *only* on the existence of the infinite substance, God, who creates it. Once it is created by God, a finite substance's existence is independent of any other substance. There are two finite substances according to Descartes: mind and body. He argues that mind can exist even if there are no bodies, and vice versa.

This, in turn, gave rise to a debate among rationalist philosophers of the day over how these two distinct substances, mind and body, causally interact with each other. How, for example, can my wish to raise my arm be the cause of my raising my arm? Given Descartes's view that mind and body are separate, this is difficult to answer. He himself ventured that there is a part of the brain that somehow allows the mind to command the body. However, this seems to contradict his insistence on the separateness of mind and body—it appears to suggest the existence of a bodily thing that is mental. Some Cartesian philosophers had attempted to solve the problem by denying that there is any causal interaction between mind and body, a view that Spinoza also defended.

---

1. Benedictus de Spinoza, *Ethics*, in *A Spinoza Reader: The Ethics and Other Works*, trans. Edwin Curley (Princeton: Princeton University Press, 1996), 146.

MODULE 4
THE AUTHOR'S CONTRIBUTION

KEY POINTS

- *Ethics* teaches that in order to live well in a universe with no place for free will, human beings need to understand our own psychological states and to discipline them.

- Spinoza felt the precision of his philosophy could best be expressed through Euclid's* geometry. He defined God as "the absolute infinite."

- Spinoza's *Ethics* is based on the rationalist* framework introduced by René Descartes.* But while Descartes's model allows people free will, Spinoza's determinism* does not.

Author's Aims

The fundamental aim of Benedictus de Spinoza's *Ethics* is to point the way to human well-being in a deterministic world that is indifferent to our concerns. The way to do this is by coming to a clear understanding of the causes of events, material and mental, and thus to reach the clearest possible comprehension of the world. An account of metaphysics* and epistemology* (the branches of philosophy that examine the concept of being, and the nature and limits of knowledge, respectively) serves as a foundation for Spinoza's psychology, which in turn leads to his theory of action and his overall account of how to live a good life. Thus the final aim of offering an account of the good life involves dealing with serious issues in the fields mentioned above. *Ethics* first tackles the most abstract questions, then moves systematically toward the particulars.

And so *Ethics* begins with a basic questioning of the nature of substance. Here Spinoza intends to clear up several problems found in previous philosophers' accounts of the topic, particularly that of René Descartes, whose concept of substance led him to claim that mind and body are two completely distinct substances. This, in turn, raised the difficult question of how minds and bodies interact, with one causing an effect in the other. Spinoza seeks to overcome the issue by arguing that mind and body can have no causal interaction (one cannot bring about action or changes in the other). *Ethics* includes an intriguing reformulation of Descartes's problem, one consequence of which is that Spinoza embraces determinism.

Spinoza extends his redefinition of substance, and his response to Descartes's problem, into the realm of epistemology. He aims to offer a rationalist account of knowledge (in other words, to claim that knowledge acquired through the senses is somehow incomplete in comparison to the mind's knowledge of itself). This underpins Spinoza's psychology, which attempts to show how our mind's knowledge of the causes of our emotions is the basis for living the good life in a deterministic universe.

> *"But I think I have shown clearly enough ... that from God's supreme power, or infinite nature, infinitely many things in infinitely many modes, that is, all things, have necessarily flowed, or always follow, by the same necessity and in the same way as from the nature of a triangle it follows, from eternity and to eternity, that its three angles are equal to two right angles."*
>
> —— Benedictus de Spinoza, *Ethics*

Approach

Spinoza placed a high value on disciplined philosophical expression, which he felt could be best achieved through mathematics. He valued Euclid's geometry* as the ideal way to express philosophical ideas, and aimed to "consider human actions and appetites just as if it were a question of lines, planes, and bodies."[1] He does this by proceeding in his writing in the manner of geometrical texts, starting with a set of definitions of his terms. For example, he defines God as, "a being absolutely infinite, that is, a substance consisting of an infinity of attributes, of which each one expresses an eternal and infinite essence."[2] This definition was, in turn, based on previous definitions of "substance" and "attributes." Having defined his basic terms, Spinoza then goes on to offer axioms* (self-evident claims), such as, "Whatever is, is either in itself or in another."[3]

As in geometrical treatises, Spinoza's axioms are to be thought of as self-evident truths that cannot themselves be proven. The axioms and definitions are the basis on which all subsequent proofs will follow. All the remaining material consists of proofs, demonstrations, and *scholia*—explanatory comments on specific demonstrations and proofs. The proofs are demonstrated either purely by reference to the definitions and axioms, or by reference to previously established proofs. Thus, as the text proceeds, it builds on earlier material.

Contribution in Context

Though Spinoza presents his definitions and axioms as general and timeless, they often contain ideas that he would claim were specifically acceptable to the philosophers of his era. Thus, for example, his definition of terms such as "substance," "attribute," and "mode" rely on Descartes's use of these terms.[4] Spinoza adopts the vocabulary of Cartesian philosophy but revises some of its basic concepts in order to address some of the fundamental problems posed by Descartes. In particular, Spinoza uses the Cartesian definition of "substance" to argue that there is an inconsistency, or contradiction, in Descartes's own use of the term. Spinoza argues, first, that if one follows the Cartesian definition of substance,[5] there cannot be more than one substance with any given attribute and,[6] second, that a substance can be produced only by another substance of the same kind.[7] These claims have radical consequences. The boldest are that substances cannot be produced and are therefore eternal[8] and without purpose.[9] Moreover, God must be the only substance and consist of infinite attributes,[10] and everything else must be a mode—or expression in another form—of God.[11]

Though starting off with assumptions common among rationalist philosophers of his day, Spinoza reaches a view that is quite remote from their outlook. In particular, his conclusion that there can be only one substance undoes Descartes's attempt to save the notion of free will from the determinism that comes with modern science. Descartes thought that determinism applies only to the material substance (that is, matter always obeys the laws of physics), while the mental substance is free from the laws of physics. Spinoza shows that this is based on an inconsistent use of the concept of substance, which in Spinoza's system forms the basis for determinism. In other words, for Spinoza, there is no free will, since the mind too is subject to the laws of the universe.

---

1. Benedictus de Spinoza, *Ethics*, in *A Spinoza Reader: The Ethics and Other Works*, trans. Edwin Curley (Princeton: Princeton University Press, 1996), 152.

2. Spinoza, *Ethics*, 85.

3. Spinoza, *Ethics*, 86.

4. René Descartes, *Principles of Philosophy*, trans. V. Rodger Miller and R. P. Miller (London: Reidel, 1983), part I, sections 51–6.

5. Spinoza, *Ethics*, 85.

6. Spinoza, *Ethics*, 87.

7. Spinoza, *Ethics*, 87.

8. Spinoza, *Ethics*, 88–90, 100.

9. Spinoza, *Ethics*, 97.

10. Spinoza, *Ethics*, 90.

11. Spinoza, *Ethics*, 94.

SECTION 2
IDEAS

MAIN IDEAS

KEY POINTS

* Spinoza differs from Descartes* in putting forward the idea of "substance monism,"* the belief that there exists only one "substance," which he identifies as "God or nature."

* *Ethics* opens with an examination of "substance." Spinoza presents a discussion of the two key categories that define substance: attribute and mode.

* The book uses a framework based on Euclidean* geometry,* whose cold rational precision reflects Spinoza's idea of a universe that is indifferent to human beings.

Key Themes

Benedictus de Spinoza's *Ethics* begins with a definition of some basic metaphysical terms such as "substance," "attribute," and "mode." Based on these definitions, Spinoza argues in favor of "substance monism," the notion that there can exist only one substance, which he identifies as "God or nature."[1] Spinoza seems to be claiming here that God and nature are one and the same "substance," a view that is called pantheism.* This one substance has two fundamental qualities, which, following Descartes, Spinoza identifies as mind and body. He further argues that these two qualities exist in parallel, and that every event is a mode (or different expression) of the one substance, and has both a mental and a bodily expression.

The text consists of five mutually reinforcing sections. It

argues that human freedom lies in the exercise of the intellect, while clarifying the nature of knowing, the nature of being, and God. Given their centrality in *Ethics'* argument, these main metaphysical concepts are fully developed in the book. The sections in *Ethics* on epistemology* and theory of action are considered to be equally important.

> *"I shall consider human actions and appetites just as if it were a question of lines, planes, and bodies."*
>
> —— Benedictus de Spinoza, *Ethics*

Exploring the Ideas

Spinoza opens *Ethics* with an argument regarding the nature of substance, which he defines, under the influence of Descartes, as "what is in itself and conceived through itself."[2] In other words, Spinoza holds that a "substance" is that kind of entity that can exist independently of other entities (one whose existence is not determined by, and does not presuppose, the existence of another kind of entity).

Beginning with this definition, Spinoza draws a number of conclusions about the nature of substance. He rejects the distinction between what Descartes had called "finite substances," that is, substances (*res cogitans** and *res extensa**) that exist independently of each other, but are created by an "infinite substance" (Descartes's God). According to Spinoza, there can be only one substance, which he famously referred to as God or nature (*Deus sive Natura*).

As we have seen, Spinoza's system looks at the category of substance. This is a crucial ontological* factor (that is, it relates to the study of being). Spinoza also employs two other categories: the "attributes" of the one substance and the "modes" of the one substance. At the start of *Ethics*, he defines the category of attribute as "what the intellect perceives of a substance, as constituting its essence."[3] It is difficult to make out exactly what Spinoza means by this unclear definition, which has been interpreted in conflicting ways. As in the case of substance, he seems to build on Descartes's concept of attribute, according to which the key attribute of mind is thought, while the key attribute of the body is extension—in other words, a body is, at its most basic, the kind of thing that is extended in physical space. Spinoza went against Descartes, however, when he argued that mind and body cannot be, as Descartes thought, separate substances; rather he maintained that mind and body are *attributes* of the one substance. For Spinoza, thought and extension are God's thought and God's extension (that is, they are essential attributes of the one substance, which is God). Somewhat confusingly, Spinoza seems to suggest that there are an infinite number of other ways in which substance manifests itself—human beings happen to have access only to these two.[4]

The third basic ontological category we find in Spinoza— along with substance and attribute—is that of mode. This he defined as "the affections of a substance, or that which is in another through which it is also conceived."[5] This definition raises the same problems of interpretation as those found in the definition of attribute, since according to this, modes are conceived through attributes. Any individual thought is a mode that can be classified under the general attribute of thought, and any individual body is a mode that belongs to the general attribute of extension. Infinite modes include the *general* laws governing all of thought, and the *general* principles governing all extended bodies (that is, the laws of geometry and physics). Finite modes, on the other hand, include all *particular* thoughts and bodies.

Spinoza's ontological scheme is the basis for his approach to Descartes's problem regarding the relation between mind and body. According to Spinoza, mind and body are simply two aspects of the same underlying reality, namely the one substance (God or nature). Mind and body are, Spinoza claims, two parallel ways in which this substance expresses itself: each event has a bodily and a mental aspect. For Spinoza, there is a mental event parallel to every bodily event, even if no human mind thinks it: all mental and bodily events are modes of the one substance (they occur in the mind and body of God or nature). Spinoza can thus go on to develop an account of human knowledge and action explained in terms of the relations held by human beings to the one substance's attributes of mind and body.

Language and Expression

Ethics unfolds within a framework explicitly borrowed from Euclidean geometry, which Spinoza valued for its precision, consistency, and universality. About 80 percent of the text of *Ethics* consists of an elaborate matrix of definitions and propositions that are self-referential—one definition or proposition often refers to others in the text. Yet each of the work's five chapters also includes textual passages, the so-called "*scholia*," which were written in a more accessible voice. Indeed, one senses in these passages a sort of detachment, irony, and even humor on Spinoza's part. This helps considerably to bring unity to many of the work's seemingly distinct projects. The question of the relationship between these two styles of expression—rigid mathematics and subtle irony—goes some distance toward making the content of *Ethics* as a whole more understandable. The claims of *Ethics* cannot be separated from the form in which they are presented. The eternal, changeless, and rational nature of geometric forms, which are indifferent to human purposes, supports Spinoza's view of the universe as purposeless.

Unfortunately for most readers, the mathematical style makes *Ethics* difficult to understand. Its vocabulary has been called strange, and its themes are extremely complicated, even impenetrable.[6] In spite of its difficulty, form and content combine in *Ethics* to generate a powerful set of conclusions for both public and academic life. All of this makes *Ethics* particularly useful when reading the history of philosophy is a source of cultural understanding in the present.

---

1. Benedictus de Spinoza, *Ethics*, in *A Spinoza Reader: The Ethics and Other Works*, trans. Edwin Curley (Princeton: Princeton University Press, 1996), 97–100.
2. Spinoza, *Ethics*, 85.

3. Spinoza, *Ethics*, 85.
4. See: Abraham Wolf, "Spinoza's Conception of Attributes," in *Studies in Spinoza, Critical and Interpretive Essays*, ed. S. Paul Kashap (Berkeley, CA: University of California Press, 1972), 26.
5. Spinoza, *Ethics*, 85.
6. See: Beth Lord, *Spinoza Beyond Philosophy* (Edinburgh: Edinburgh University Press, 2012), 1.

SECONDARY IDEAS

KEY POINTS

- Spinoza goes from some very abstract metaphysical* claims to some particular assertions on how one can live a good life. He does this by giving an account of the nature of knowledge and of the emotions.

- In his epistemology*, psychology, and theory of action, Spinoza demonstrates the implications of some of his most influential and perhaps controversial doctrines, such as his determinism.*

- While a lot of attention has been paid to Spinoza's metaphysics, his particular ideas about ethics have sometimes been overlooked.

Other Ideas

In *Ethics*, Benedictus de Spinoza's metaphysics (the branch of philosophy examining the nature of being) is intertwined with his epistemology (philosophical investigation into the nature of knowledge). The latter is largely an extension of his view of the parallelism between mind and body. Spinoza argues that the individual mind, rather than being a storehouse of private mental content separate from an outer world, is itself a part of the world that is closely bound up with the body; in a sense, they are two sides of the same coin. Knowledge is an essential part of the intellect. Spinoza here makes a distinction between knowledge acquired through the senses ("inadequate ideas") and knowledge acquired through either reason or intellectual intuition ("adequate ideas"). Our senses and imagination give us "confused and mutilated"[1] mental pictures of material things, which are partial and subjective. In contrast, Spinoza claims that reason and intellectual intuition (*scientia intuitiva**) can lead us to knowledge of timeless truths. Here, Spinoza seems to be rejecting the empiricism*— which claims that all knowledge is acquired through the senses and denies the existence of knowledge acquired solely through reason or intellectual intuition—developed by his contemporary John Locke,*[2] with his emphasis on reason. Spinoza thus placed himself in the camp of the rationalists* (his relation with empiricism is, however, a matter of debate among scholars).

Spinoza's epistemology is applied to his psychology of the emotions, and to his theory of action. At the heart of these lies the *conatus** principle: "Each thing, as far as it can by its own power, strives to persevere in being."[3] Spinoza states this in the context of his discussion of the emotions, or "affects." He sees the affects as coming from joy, sadness, and desire—the three basic ways upon which the mind is acted. Joy is the result of an increase in the power to persevere in being, while sadness is the opposite; desire is the striving toward perseverance itself (that is, the effort to remain alive). All other affects are formed through combinations of these elements. For example, as Spinoza puts it, "Hope is an inconstant joy, born of the idea of a future or past thing whose outcome we to some extent doubt."[4]

Spinoza makes a distinction between passive and active affects. Joy and sadness are considered passive affects, as they are "inadequate ideas," or ways in which the mind is influenced by factors external to it. Joy and sadness can, however, be transformed into active affects once the mind grasps them clearly.[5] This means having an intellectual intuition of the affects as being the results of a causal chain that deterministically leads to them.

This takes us to Spinoza's theory of action, which relies on the argument that one acts rightly when aided by the intellect's recognition that free will is an illusion. Freedom, according to Spinoza, comes precisely from the realization that our actions are the results of a chain of cause and effect. In other words, freedom is not freedom of the will, which is illusory. Our language of good and evil has relied on a false conception of free will, which Spinoza's ethics revises on the basis of the *conatus* principle. The mind strives to persevere by turning affects from passive to active and understanding its own nature.

> *"By God I understand a being absolutely infinite, that is, a substance consisting of an infinity of attributes, of which each one expresses an eternal and infinite essence."*
>
> —— Benedictus de Spinoza, *Ethics*

Exploring the Ideas

One of the basic ideas that takes us from the abstract metaphysical claims at the beginning of *Ethics* to its particular psychological and ethical claims is that of intellectual intuition (*scientia intuitiva**). Intellectual intuition is a third kind of knowledge beyond the distinction between reason and knowledge gained through the senses. Spinoza vaguely defines it as knowledge that "proceeds from an adequate idea of the formal essence of certain attributes of God to the adequate knowledge of the formal essence of things."[6]

By intellectual intuition, Spinoza means grasping or intuiting that a thing follows necessarily from the nature of the one substance. This is neither knowledge acquired through reason nor through the senses, but is a third type of direct intuitive knowledge. In other words, it serves the crucial purpose of being the state of knowing in which the deterministic nature of the universe is glimpsed.

Spinoza's account of intellectual intuition relies on a crucial earlier step in his metaphysical system, that of distinguishing between *natura naturans** and *natura naturata*,* translated as "nature naturing" and "nature natured." This is a distinction between the "active" principle that produces nature, and its "passive" product. *Natura naturans* is the cause, while *natura naturata* is its necessary effect. Intellectual intuition functions as a means of granting the intellect direct knowledge of the whole of nature, albeit only as *natura naturata*. Intellectual intuition grasps the passive product, not the active cause of nature. It is a recognition that the specific thing under scrutiny follows necessarily from the nature of God—that it could not have been otherwise. Knowing things in this way is the basis of freedom from the bondage of passive affects.

Overlooked

Ironically for a work whose title is *Ethics*, it is the book's ethical content that is most overlooked. In the text's initially harsh reception, when Spinoza was charged with atheism* and his ideas were rejected across Europe, it was his views on God that received the most attention. Even a century after the book's publication, during the German pantheism controversy* of the 1780s, which opened up *Ethics* to broader critical attention, the claims on God and nature in parts one and two rather than the ethics in parts three through five stood at the center of the discussion.[7] The consequences of this focus were that Spinoza came to be read as a metaphysician arguing from a particular tradition: pantheism*[8]— a tradition that was unacceptable to the established religions. A more charitable interpretation, however, would see Spinoza as seeking to go beyond tradition entirely, directing his project toward understanding human freedom. Focusing on metaphysics as a particular tradition at the expense of the book's insights into ethics does not do the work justice. It hides the sense in which all of its parts are intended to guide the reader toward Spinoza's critical inquiry into how one can come to live the good life, which Spinoza sees as a life of blessedness.

1. Benedictus de Spinoza, *Ethics*, in *A Spinoza Reader: The Ethics and Other Works*, trans. Edwin Curley (Princeton: Princeton University Press, 1996), 135.

2. See: Alexander Douglas, "Was Spinoza a Naturalist?" *Pacific Philosophical Quarterly* 96, no.1 (2015): 85.

3. Spinoza, *Ethics*, 159.

4. Spinoza, *Ethics*, 190.

5. Spinoza, *Ethics*, 247.

6. Spinoza, *Ethics*, 141.

7. See: Friedrich Heinrich Jacobi and Gérard Vallée, *The Spinoza Conversations Between Lessing and Jacobi: Text with Excerpts from the Ensuing Controversy* (University Press of America, 1988).

8. See: John Dewey, "The Pantheism of Spinoza," *The Journal of Speculative Philosophy* 16, no. 3 (1882): 249–57; F. C. S. J. Copleston, "Pantheism in Spinoza and the German Idealists," *Philosophy* 21, no.78 (1946): 42–56.

MODULE 7
ACHIEVEMENT

KEY POINTS

* Spinoza's *Ethics* constructs an elaborate system of proofs in a geometrical style, creating a highly interconnected system. Elegant as it is, it has the disadvantage that if one proof or definition is rejected, the whole work unravels.

* *Ethics*, today considered one of the most important works of modern philosophy, has influenced other great writers. Yet when it was published, it was seen as denying the existence of God, a grave scandal in the seventeenth century.

* Despite the fact that the highly interconnected structure of *Ethics* seems to come undone if any of its parts is rejected, the work is highly innovative and contains keen observations about human life.

Assessing the Argument

Benedictus de Spinoza's grand task in *Ethics* is to start out from a number of abstract axioms* (self-evident claims) and definitions, and through a series of proofs present a thorough view of God, human knowledge, the psychology of affects, and freedom. The sheer breadth of this task can be intimidating to the reader, as can the precision with which Spinoza undertakes it. The text builds on itself, with each new proof relying on earlier definitions, axioms, and proofs. At times it feels like a maze: Spinoza continually refers back to earlier proofs, in the geometrical style that he adopts in his writing.

    Spinoza's geometrical method creates a system with one

big disadvantage: if one were to reject some evidence, axiom, or definition in *Ethics*, all of the other elements would have to be rejected as well. Due to its argumentative structure—the geometrical method—the text's grand systematic achievements can be dismissed by rejecting details, without paying careful attention to the entirety of its system, and there are many good reasons for rejecting a number of Spinoza's views. For example, his ambitious task includes an attempt to prove God's existence. Though proving God's existence was a key part of philosophy at the time, later developments in the history of philosophy raise questions about such efforts. Further doubts remain about the internal logic of Spinoza's arguments, since it is not clear that accepting all of his definitions and axioms leads to an acceptance of all his proofs. He sometimes relies on *scholia** or new definitions placed in the text to introduce views that could not be proven by earlier definitions or axioms.

Spinoza's conclusions are often more interesting than the formal reasoning through which they are obtained, and may merit reworking by contemporary philosophers.

"You are either a Spinozist or not a philosopher at all."
—— G. W. F. Hegel, *Lectures on the History of Philosophy*

Achievement in Context

Given the difficulties the book presents to the reader, it is possible to misunderstand its more controversial views, such as those regarding the nature of God. For a century or so after its publication, for example, the theology of *Ethics* was interpreted as being atheistic,* or denying the existence of God. Later interpreters, however, took it to be pantheistic,* proposing that God is everywhere (a belief that clashes with most monotheistic religions).

Spinoza's thoughts on religion were considered radical. This meant his views were seen as a kind of heresy* in philosophical circles: being accused of "Spinozism" was taken to be an insult against which one ought to defend oneself. Yet these attacks on Spinoza's thought came more from a cultural bias rather than from a balanced analysis of his philosophical arguments. The context in which he was writing would not have accepted as normal the idea that the word "God" might be followed by the phrase "or nature." Similarly, most philosophers would have been upset with Spinoza's arguments against anthropomorphic* conceptions of God.

Ethics remains one of the most important works of philosophy ever written. This is because of the range of concerns it deals with, covering everything from the nature of being to the mind's relationship to the body and the ways to achieve happiness in a deterministic* world. These topics go beyond philosophy itself, extending into cognitive therapy (a type of psychotherapy), environmentalism, politics, and art. For example, the Argentine author Jorge Luis Borges,* whose work blends literature, fantasy, and philosophy, claimed a considerable debt to Spinoza.[1] Likewise, Johann Wolfgang von Goethe,* one of the most famous German writers of the eighteenth and early nineteenth centuries, praised Spinoza for helping him calm his excessive passions as a young man.[2] In addition, Spinoza's work has been seen as bridging the distinction between Eastern and Western thought.

The style of *Ethics* is perhaps another reason why it is seen as so widely important. At the cost of ease of understanding, Spinoza deliberately selects terminology from mathematics as a means not only to express his points as precisely as possible, but also to communicate them in a way that is equally meaningful to all cultures. To be sure, Spinoza does not always succeed in this aim, and *Ethics* contains language that reflects the work's various historical influences. These include Spinoza's background in Marrano* Jewish thought, the Enlightenment* rationalism* of René Descartes,* and even the mathematical style itself, which is rooted in Euclid's* *Elements*.

Limitations

Spinoza borrows the style of Euclidean geometry*—for instance, the assumption that there is something intuitively obvious about the idea that two parallel lines extended to infinity will never cross. A similar case can be made about intuitively obvious logical principles, such as the principle of identity (A=A). No proof can be offered of this principle, though it can be thought to be self-evident for all rational beings. The question is whether there is any natural reason for the reader to accept Spinoza's basic assumptions. In contrast with the intuitive nature of the principles of logic or the axioms of geometry, it seems it might take some more work to accept Spinoza's complex and unclear starting points. This risks upsetting the balance between abstract definitions and relatively more concrete claims—which is the structure of each of the work's five sections.[3] For if the more concrete claims depend for their truth on their more abstract counterparts, then rejecting those abstract parts also means rejecting all that is concrete that follows from them. Additionally, if, on the other hand, the work's abstract claims depend in any way on their more concrete counterparts, then this undermines the geometrical method, which proceeds from the abstract to the concrete. Yet despite its shortcomings, *Ethics* made important innovations in philosophy and develops a strikingly original theory of being, one with detailed observations about human life.

---

1. See: Marcelo Abadi, "Spinoza in Borges' Looking-Glass," *Studia Spinozana: An International and Interdisciplinary Series* 5 (1989): 29–42.
2. Stuart Hampshire, *Spinoza* (Manchester: Manchester University Press, 1956), 18.
3. See: Guttorm Fløistad, "Spinoza's Theory of Knowledge," *Inquiry* 12, nos.1–4 (1969): 41–65.

PLACE IN THE AUTHOR'S WORK

KEY POINTS

- Spinoza's earlier works, and his surviving correspondence with a number of people, provide insights into the development of his ideas.

- *Ethics* deals with many of the issues raised by Descartes* and his followers, such as the relation between mind and body. Spinoza's powerful writing played a big role in promoting the scholarly study of the Bible.

- Spinoza's excommunication,* and a biographical entry on him written shortly after his death, left him with a bad reputation for the next century.

Positioning

Ethics was Benedictus de Spinoza's greatest work. While he did write a number of other texts, these are of interest to us as much to gain further information about *Ethics* as for any insights they themselves contain. These other works show the development and the consistency of his thinking during the two decades between his excommunication and his death. The most notable among Spinoza's other writings are his letters to a network of people with whom he corresponded, his *Treatise on the Emendation of the Intellect* (1662), *Parts One and Two of the Principles of Philosophy of René Descartes Demonstrated According to the Geometric Method* (1663), the *Theological-Political Treatise* (1670), and his unfinished *Political Treatise*. It is helpful to take these in turn.

Approximately 50 of Spinoza's personal letters remain today.[1] These date from approximately 1660 through to Spinoza's death in 1677, and aside from what they show about the author's personality and character these are of interest because they reveal that as early as 1660, he was thinking deeply about several key claims, such as the unity of God and reason, that would later appear in *Ethics*.

Also significant among Spinoza's early writings, the *Treatise on the Emendation of the Intellect* ("emendation" means "revising" or "correcting") was begun in the late 1650s and addressed the proper means to achieving true understanding.[2] Like the letters, this work seeks to distinguish clear and distinct true ideas from the inadequate ones that mislead us. The text shows Spinoza's notable debt to René Descartes. Descartes's influence on Spinoza is most on display in the latter's 1663 text, *Parts One and Two of the Principles of Philosophy of René Descartes Demonstrated According to the Geometric Method*.[3] Written at the request of friends who wished to gain a better understanding of Descartes's ideas, this work is significant because it marked a turn toward the geometric style of *Ethics*. It was also the only text that Spinoza published under his own name.

The work that comes closest to *Ethics* chronologically was his *Theological-Political Treatise*, published in 1670.[4] A sort of plain-language introduction to some of the ideas that *Ethics* would later express geometrically, the *Treatise* offers a rationalist basis for religious liberty, and was published during a break in composing *Ethics*.

A final work, one that was never finished and remains a bit of a mystery, is the *Political Treatise*.[5] Spinoza began it in the middle of 1676, after *Ethics* had been completed. Intended as a sequel to the *Theological-Political Treatise*, it aimed to show how states with different constitutions can be made to function well, and thus serve as an argument for democracy. Despite the interest attracted by these other texts, Spinoza's reputation depends mainly on his main work, *Ethics*.

> *"The* Ethics *is not only Spinoza's masterwork, it is also his life's work. We know from the correspondence that he began writing it early in the 1660s, that a substantial draft of the work was in existence by 1665, and that he then put it aside to write his* Theological-Political Treatise, *which appeared in 1670. He had published his exposition of Descartes' philosophy to pave the way for his* Ethics ... *the* Theological-Political Treatise *had a similar motivation."*
>
> —— Edwin Curley, "Spinoza's Life and Philosophy"

Integration

The whole of Spinoza's output might have seemed inconsistent if he had not published *Ethics*. The topics dealt with in his previous writings involved a clear interest in the philosophy of Descartes and his followers, which had, at the time Spinoza was writing, only shortly before come to dominate philosophical discussion in the Netherlands. On the other hand, Spinoza seems deeply concerned with theological and political issues, single-handedly prompting the secular study and interpretation of the Bible. These two interests are the subject of the two most significant works he published during his lifetime, the *Principles of Philosophy of René Descartes* and the *Theologico-Political Treatise*.

These two texts are, however, in some ways synthesized in *Ethics*. The *Principles*, for example, introduces Spinoza's geometrical method of writing about philosophy. It also forms the background of his mainly Cartesian concerns within *Ethics*, which is to a large extent a response to questions raised by Descartes and his followers, such as the relation between mind and body. *Treatise*, on the other hand, deals with themes investigated in *Ethics*, such as the rejection of anthropomorphic conceptions of God (that is, the notion of God as having human characteristics), the psychology of affect and its place in religion, and the notion of freedom.

Significance

Ethics brings Spinoza's previous concerns together in a systematic manner, using its geometrical style to attempt to demonstrate his views. The work covers many areas, ranging from metaphysics* and theology to epistemology* and psychology, all of which are interwoven.

Though *Ethics* was Spinoza's most important work, for the century following its publication, other factors shaped Spinoza's reputation as a thinker. The rumors surrounding Spinoza's life, especially Pierre Bayle's* biography of him in his *Historical and Critical Dictionary* (1697), helped to establish his bad reputation. Spinoza's excommunication early in his life, as well as the controversy surrounding the publication of the *Theologico-Political*

Treatise and his association with republican leader Jan de Witt,* gave hostile interpreters the grounds on which to read *Ethics* as a defense of atheism.* This is an uncharitable interpretation of Spinoza, especially given that he seems to hold that the highest state of being is what he describes as an intellectual love of God (*"amor dei intellectualis"*). It was only around a century after Spinoza had died, in the so-called pantheism controversy* among German intellectuals, that his reputation was reconsidered. This involved the development of a new interpretation of *Ethics*, no longer seen as backing atheism, but rather as promoting a form of pantheism (God is nature). This view may be more charitable toward Spinoza, though contemporary interpreters of his thought are still debating whether we should see him as an atheist, pantheist, or panentheist* (who sees God as *in* all nature).[6]

---

1. Benedictus de Spinoza, *The Correspondence of Spinoza* (New York: Russell & Russell, 1928).

2. Benedictus de Spinoza, "The Treatise on the Emendation of the Intellect," in *Ethics: With The Treatise on the Emendation of the Intellect and Selected Letters*, trans. Samuel Shirley (Indianapolis: Hackett Publishing, 1992), 233–62.

3. See: Benedictus de Spinoza, *The Principles of Cartesian Philosophy: With, Metaphysical Thoughts*, ed. Samuel Shirley (Indianapolis: Hackett Publishing, 1998).

4. Benedictus de Spinoza, *Theological-Political Treatise* (Cambridge: Cambridge University Press, 2007).

5. Benedictus de Spinoza, "Political Treatise," in *A Theologico-Political Treatise and A Political Treatise*, ed. Francesco Cordasco (New York: Courier Corporation, 2013), 267–388.

6. See: Genevieve Lloyd, *Routledge Philosophy GuideBook to Spinoza and the* Ethics (London: Routledge, 1996), 40.

SECTION 3
IMPACT

THE FIRST RESPONSES

KEY POINTS

- *Ethics* was at first blindly condemned for its supposed atheism* and immorality. It took a century before philosophers began to see a wealth of wisdom in the work.
- So strong was the condemnation of Spinoza that the famous philosopher Leibniz* joined in publicly—even though secretly, he was in contact with Spinoza and adopted some of his ideas.
- During the pantheism controversy,* a hundred years after Spinoza's death, the leading philosopher Gotthold Lessing* admitted that Spinoza's pantheism* had influenced him, setting the stage for an acceptance of Spinoza's work.

Criticism

In the period following its publication, the reaction to Benedictus de Spinoza's *Ethics* was hostile, thanks in large part to accusations of atheism and immorality in Pierre Bayle's* widely read *Historical and Critical Dictionary* (1695). According to the intellectual historian Peter Gay,* Bayle's dictionary "misled a whole century."[1]

The first significant thaw in Spinoza's reception came in Germany in the 1780s, arising out of a debate between Gotthold Lessing and Friedrich Heinrich Jacobi,* which came to be known as the pantheism controversy. In search of an alternative to the forms of thought then dominant in the German Enlightenment,* Lessing had confessed to finding in Spinoza's *Ethics* a system he could embrace. Although the response to Lessing's admission was not entirely favorable, the resulting discussion shifted significantly the way Spinoza was viewed. Rather than the atheist caricature of Bayle's *Dictionary*, Spinoza came to be read as the founder of his own coherent metaphysical*/ theological system: pantheism. From here, it was not long before philosophers across the German idealist* movement of the late eighteenth and early nineteenth centuries found parts of *Ethics* that appealed to them. Attention focused almost entirely on *Ethics'* metaphysics and theology (parts one and two) rather than its ethics (parts three through five). The leading German philosopher G. W. F. Hegel,* for example, argued in his *Lectures on the History of Philosophy* (1837) that Spinoza's idea of substance formed the "foundation of all true views," and that the real reason for the harsh reaction to his ideas was that the critics could not bear the thought of their own annihilation in the oneness of nature.[2] To a large extent, the pantheism debate was the start of patterns that continue to shape discussions about the place of *Ethics* in Western philosophy, from its being taken up on behalf of a variety of other agendas to the focus on the work's opening two sections—on God and nature—at the expense of its final three.

"He asserts therefore the most infamous and most monstrous extravagances that can be conceived, and much more ridiculous than the poets concerning the gods of paganism. I am surprised either that he did not see them, or if he did, that he was so opinionated as to hold on to his principle. A man of good sense would prefer to break the ground with his teeth and his nails than to cultivate as shocking and absurd a hypothesis as this."

—— Pierre Bayle, *Historical and Critical Dictionary*

Responses

Since Spinoza died before *Ethics* was published, he never got the chance to respond to critics of the work. Evidence from his letters, however, suggests that he was often dismissive of even friendly criticism from his peers.[3] Indeed, whenever Spinoza was confronted by someone with whom he was corresponding over some difficulty in his work, he would often brush the person off completely—a tendency that increased in his later years. This tendency in Spinoza's letters leads to the suspicion that his unflappable personality would have caused him to remain calm in the face of criticisms of *Ethics*. Such self-control, at any rate, had been his reaction to both his excommunication* and to the uproar surrounding his 1670s *Theological-Political Treatise*, which among Spinoza's works is the one that most resembles *Ethics* in content, if not in form. Of course, this is speculation, as is the expectation that Spinoza's unfinished *Political Treatise* would have contained responses to criticisms of *Ethics*. Given the constant nature of his thought across his adult life, however, Spinoza appears to have been very difficult to persuade.

One interesting incident in the history of philosophy in particular can shed some light on Spinoza's response to criticism. The famous scholar and philosopher Gottfried Wilhelm von Leibniz is usually seen as a leading representative of the rationalist* tradition. That tradition begins with René Descartes,* moves through Spinoza, and concludes with Leibniz. Leibniz had engaged in correspondence with Spinoza, and in 1676, after *Ethics* was completed but not yet published, even visited him. Leibniz, however, kept this visit a secret, given Spinoza's bad reputation. In the works that he published during his lifetime, Leibniz was critical of Spinoza, and in particular he contrasted his own insistence on the existence of free will with Spinoza's determinism.* He did, however, seem to integrate a lot of Spinoza's ideas into his work, and that published after his death seems to show that in private he also held deterministic views like those of Spinoza.[4] This reflects the fate, within the context of eighteenth-century philosophy, of ideas associated with Spinoza.

Conflict and Consensus

Throughout the eighteenth century, philosophers overwhelmingly condemned Spinoza's work. Influenced in large part by the French philosopher Pierre Bayle's account of his life, Spinoza's work was hastily and unfairly seen as the work of an atheist, a man who had been condemned for heresy* by Amsterdam's synagogue, and whose *Theologico-Political Treatise* was banned by the Synod, or ruling council, of the Dutch Reformed Church. This meant that few scholars took a fair-minded view of Spinoza's work. One notable exception is to be found in Leibniz's private papers. These, however, would remain a secret throughout Leibniz's life; despite privately developing a deterministic world-view, Leibniz publically denounced Spinozist determinism, contrasting it to his metaphysics, which, he claimed, was a defense of free will.

Perhaps more than any thinker throughout the history of philosophy, Spinoza was radical enough that it took a century of scorn before his reputation began to be restored. His supposed atheism, combined with his radical, deterministic outlook, ensured that the widespread view among philosophers was of Spinozism as a kind of philosophical sin. It was only in the pantheism controversy, around a century after Spinoza's death, that a philosopher would dare to claim that Spinoza had influenced him.

---

1. Peter Gay, *The Enlightenment: A Comprehensive Anthology* (New York: Simon and Schuster, 1973), 293.

2. Genevieve Lloyd, *Routledge Philosophy GuideBook to Spinoza and the* Ethics (London: Routledge, 1996), 16.

3. See: Benedictus de Spinoza, "Objections and Replies," in *A Spinoza Reader: The Ethics and Other Works*, trans. Edwin Curley (Princeton: Princeton University Press, 1996), 146.

4. See: Bertrand Russell, *History of Western Philosophy* (Oxon: Routledge, 2004)

MODULE 10
THE EVOLVING DEBATE

KEY POINTS

* More liberalizing attitudes in the nineteenth century helped turn philosophers' attitudes toward Spinoza from rejection to deep praise. Later, even great thinkers in other fields, such as Sigmund Freud* and Albert Einstein,* spoke of Spinoza's deep impact on them.

* Few scholars have become "orthodox" Spinozists in the sense of having adopted his whole project. But even today, the work of some scholars in various fields is sufficiently influenced by the philosopher that they can be termed Spinozists.

* Spinoza's *Ethics* is still being studied not only as a valuable contribution to modern metaphysics,* epistemology,* psychology, and ethics, but also as a text that can be linked to contemporary debates in philosophy of mind, radical political thought, ethology,* and psychoanalysis.

Uses and Problems

In the nearly three and a half centuries since Benedictus de Spinoza's *Ethics* was published, general hostility toward the work has given way to widespread admiration, a transition helped by the weakening consensus on religion, a liberalization of politics, and a shifting of philosophical interests. The near-universal rejection of Spinoza began to be overturned in Germany in the 1780s with the pantheism controversy,* which started with Gotthold Lessing's* admission to being a follower of Spinoza. The resulting debates gave rise to the first positive references to Spinoza. G. W. F. Hegel* went so far as to claim that reading Spinoza formed the starting point for all philosophy.[1]

According to Spinoza scholar Pierre-François Moreau,* the pantheism* debate marked the end of the Enlightenment,* helped to create the artistic and literary movement called Romanticism,* and brought respectability to Spinoza's thought.[2] Whatever the merits of Moreau's position, it is clear that ideas in philosophical circles had significantly shifted by the nineteenth century toward a more favorable reception of *Ethics*. For example, changing attitudes to atheism were such that Friedrich Nietzsche* saw in Spinoza a precursor.

In the twentieth century, Sigmund Freud spoke of his high regard for Spinoza, and the twentieth-century philosopher Bertrand Russell* seems to have had a broadly speaking Spinozist view of ethics. Albert Einstein also thought highly of Spinoza. When pressed by a rabbi, Einstein claimed he believed in Spinoza's God, a deity that is rational and indifferent to human concerns. Interest in Spinoza is perhaps stronger now than it has ever been, and historical, textual, and philosophical work on him continues to appear at a rapid rate.

> "Spinoza is the Christ of philosophers, and the greatest philosophers are hardly more than apostles who distance themselves from or draw near to this mystery."
> —— Gilles Deleuze and Felix Guattari,* *What is Philosophy?*

Schools of Thought

Several contemporary scholars work in a framework that can be considered Spinozist. Thinkers such as Edwin Curley* and Jonathan Bennett* have highlighted Spinoza's insights as sources for their own efforts to develop the philosophy of mind and to practice logical analysis with rigor and precision. Curley has presented Spinoza's analysis of the relation between thought and extension as properties of God,[3] and Bennett's *A Study of Spinoza's Ethics* draws many parallels with current debates in philosophy of mind and metaphysics.[4]

The twentieth-century French philosopher Gilles Deleuze* has combined cognitive therapy with radical politics to argue that reading Spinoza helps open the reader to new ways of thinking about current social and political issues.[5] Such insights have served to ground Deleuze's burgeoning discipline of ethology, which looks at human and social behavior from a biological point of view.

Akin to Deleuze's analysis, a small but growing strand of commentators maintain that one must understand Spinoza in a fuller and more complete historical context.[6] Robert S. Corrington* has applied Spinoza's ideas, particularly the distinction between *natura naturata** and *natura naturans*,* toward what he calls "ecstatic naturalism." This combines Spinoza's naturalism* with the German idealist* philosopher Friedrich Schelling's* emphasis on spontaneity within nature, as well as elements of the pragmatic philosophical tradition, to comment upon the experiences of the sacred within nature.[7] There are also Spinoza disciples who are not primarily philosophers, as was the case with Johann Wolfgang von Goethe,*[8] Jorge Luis Borges,*[9] and Sigmund Freud.[10] It is important to bear in mind, however, that among these figures, virtually none applies Spinoza's work as a whole to any project.

In Current Scholarship

Amid widespread agreement as to Spinoza's importance, there are nevertheless many different—even opposing—uses to which his thinking has been put. Deleuze,[11] for example, controversially reads Spinoza as putting forth a brand of "empiricism." In Deleuze's work, this stands as a rejection of the existence of transcendent entities, as found in Spinoza's identification of God with nature. For Deleuze, Spinoza's understanding that the mind is equally subject to determinism* as the body has certain political connotations. It undoes the Cartesian insistence on the solitary nature of subjectivity, showing the mind to be a part of the world, involved in relations that determine its states. Deleuze's approach to Spinoza focuses on the practical implications of his philosophy, seeing Spinoza as integrating the theoretical discipline of ontology* with the practical discipline of ethics.

Many other current thinkers have instead focused only on *Ethics*' more theoretical aspects.[12] For example, there is an open debate concerning the relationship between logic and psychology in the text. Bennett has argued that Spinoza fails to properly distinguish logic and psychology,[13] whereas Albert Balz* argued that Spinoza's account of logic excludes psychology entirely.[14] Another ongoing discussion concerns Spinoza's concept of causality, upon which hinges the question of causality between mind and body. Charles Jarrett has claimed that *Ethics* argues that no mental event can cause a physical event, and vice versa.[15]

However, commentators such as Donald Davidson and Olli Koistinen disagree.[16] These authors hold that Spinoza's denial of causal interaction can be countered through attention to what they call "transparent" causality, in which it is always true that if X causes Y and X is identical with Z, then Z causes Y. Thus, if a physical change in your brain causes a muscle in your arm to contract, and that brain change is a mental decision, then the decision causes your muscle contraction.

1. See: G. W. F. Hegel, *Lectures on the History of Philosophy, Volume 3: Medieval and Modern Philosophy*, trans. E. S. Haldane and Frances H. Simson (Lincoln: University of Nebraska Press, 1995), section 2, chapters 1, A, 2.

2. Pierre-François Moreau, *Spinoza: L'expérience et l'éternité* (Paris: Presses Universitaires de France, 1994), 420–1.

3. Edwin Curley, *Spinoza's Metaphysics: An Essay in Interpretation* (Cambridge, MA: Harvard University Press, 1969).

4. Jonathan Bennett, *A Study of Spinoza's Ethics* (New York: Hackett, 1984).

5. Gilles Deleuze, *Expressionism in Philosophy: Spinoza* (Cambridge, MA: MIT Press, 1990).

6. See: Willi Goetschel, *Spinoza's Modernity: Mendelssohn, Lessing, and Heine* (Madison, WI: University of Wisconsin Press, 2004).

7. Robert S. Corrington, *Ecstatic Naturalism: Signs of the World* (Indianapolis: Indiana University Press, 1994).

8. Stuart Hampshire, *Spinoza* (Manchester: Manchester University Press, 1956), 18.

9. See: Marcelo Abadi, "Spinoza in Borges' Looking-Glass," *Studia Spinozana: An International and Interdisciplinary Series* 5 (1989): 29–42.

10. See: Walter Bernard, "Freud and Spinoza," *Psychiatry* 9, no. 2 (1946): 99–108.

11. Gilles Deleuze, *Expressionism in Philosophy: Spinoza* (Cambridge, MA, Mit Press, 1990); Gilles Deleuze, *Spinoza: Practical Philosophy*, trans. Robert Hurley (San Francisco: City Lights Books, 1988).

12. See, for instance: Jonathan Bennett, *A Study of Spinoza's* Ethics (New York: Hackett, 1984).

13. Bennett, *A Study*.

14. Albert G. A. Balz, *Idea and Essence in the Philosophies of Hobbes and Spinoza* (New York: Columbia University Press, 1918).

15. Charles E. Jarrett, *Spinoza: A Guide for the Perplexed* (New York: Continuum, 2007).

16. Donald Davidson, "Spinoza's Causal Theory of the Affects," in *Desire and Affect: Spinoza as Psychologist*, ed. Yirmiyahu Yovel (Leiden: Brill, 1999), 95–111; Olli Koistinen, "Causality, Intentionality, and Identity: Mind–Body Interaction," in *Ratio* 9, no.1 (1996): 23–38.

MODULE 11
IMPACT AND INFLUENCE TODAY

KEY POINTS

• Since the pantheism controversy* of the 1780s made it acceptable for philosophers to discuss the rich contributions of *Ethics*, the book has inspired thinking in a wide range of disciplines.

• Spinoza's aim in *Ethics* was to discredit teleological* thinking (which examines the purpose of things), the belief in freedom of the will, and the idea of God as separate from the world. The work remains radical even today.

• Recent philosophical responses to Spinoza have focused not only on his metaphysics* and philosophy of mind, but also on his politics, and even his ecological thinking.

Position

Benedictus de Spinoza's *Ethics* has remained relevant since it was first published. Spinoza's importance has grown as the concerns of philosophers have shifted in the past three centuries toward perspectives more receptive to his claims. This is so, for example, when modern science seeks to identify laws that produce regularities in nature that also apply to human beings. It is also the case in radical critiques of teleological thinking or of hierarchies in politics. In short, thinkers of many stripes have found in *Ethics* a valued source of inspiration.

Interest in the work has continued across different fields.[1] Some observers focus on the continuing relevance of Spinoza's attempts to wrestle with problems in metaphysics, logic, or the philosophy of mind.[2] Others have admired Spinoza for the political implications of his thinking, his efforts to trace the limits of thought, or the positive impact his thinking has had on ecology— removing as it does any special status for human beings within nature.[3]

Changes in religious belief are also responsible for the rise in Spinoza's fortunes. The reception of *Ethics* was plagued by charges of atheism* for nearly a century after its publication—a period in which expressing atheist beliefs could amount to a death sentence. However, there was a clear change in the response to the work after the debate over pantheism* among German thinkers in the 1780s widened the range of acceptable responses to Spinoza's thought. By the later nineteenth century, such an anti-clerical (opposed to the power of priests) figure as Friedrich Nietzsche* could openly embrace Spinoza as an inspiration in denying free will, or in overcoming the dichotomy between good and evil as existing within the nature of being. References to Spinoza abound throughout Nietzsche's oeuvre, and range from him enthusiastically embracing Spinoza as his precursor to more critical remarks.[4] Likewise, Spinoza's support for naturalism* (which sees the laws of nature as governing the universe) have come to be seen as pioneering for many in the modern scientific community who consider themselves naturalists.[5]

⌐ *"Some lovers of Spinoza may find this kind of inquiry, as pursued by Bennett, offensive. He does not give Spinoza high marks for deductive rigor ... But we would do no honor*

138

to Spinoza's memory if we did not recognize that Bennett is approaching Spinoza in the spirit in which he would have wished to be approached. Spinoza would have had no patience with anyone who rejected the propriety of the kind of logical questions Bennett raises."

—— Edwin Curley, "On Bennett's Spinoza:
The Issue of Teleology"

Interaction

Mild-mannered in character, Spinoza nevertheless found himself at the center of controversy at many points in his life. Most notable was his excommunication* from Amsterdam's Marrano* Jewish community at the age of 24 as a result of the radicalism of his ideas and his refusal to compromise what he saw as the philosopher's duty to seek rational explanations. Thus, *Ethics*, while not written expressly as a counterattack, nonetheless presented a radical, direct, conscious challenge to many forms of established thought. With the aim of discrediting all forms of teleological thinking (seeing a purpose in the cause of all things), as well as the belief in freedom of the will, and in God as being separate from the world, *Ethics* was radical in its time and remains so today. Indeed, Spinoza's democratic impulse challenges the traditional idea of order and hierarchy to its core. Spinoza has been seen as a very early proponent of environmentalism.[6]

Ethics presents a way of doing philosophy that interacts with other disciplines (for example, psychology and politics). Yet it cannot be reduced in its methods to any particular style other than the geometrical method.* In other words, form is essential to the content of the work, part of what makes it so unique and lasting, even if it is often confusing for the reader.

One of the most significant legacies of *Ethics*, one highlighted by such figures as Jonathan Bennett,* is that the work's failures are instructive, and at least force one to respond to the failures to provide a better alternative.[7]

The Continuing Debate

Much of *Ethics*' legacy lies in the fact that it makes an innovative case for the rules that govern nature and God alike to be eternal. Thanks to its complexity and range of applications, this argument has inspired a wide set of projects. Many of these projects have of course had their critics. Yet regarding *Ethics* itself, there simply are not many thinkers today whose primary aim is to criticize Spinoza. This is not to say that Spinoza's ideas are currently beyond reproach, much less that *Ethics* is no longer of interest. The philosopher of religion Martijn Buijs, for instance, argued recently in his paper "How to Make a Living God" that Spinoza's account of freedom suffers from its refusal to acknowledge the element of chance. In this respect, Spinoza is often compared unfavorably to the more open metaphysics of the German idealist* philosopher Friedrich Schelling,*[8] who wrote a century and a half after Spinoza, in the first half of the nineteenth century. The point, though, is that Spinoza's thought has been absorbed widely and deeply into the modern philosophical landscape. As such, most of the contemporary approaches to his work mentioned so far harness the innovations of *Ethics* rather than challenging it wholesale. Or, to put it another way, the text is simply so unique that criticisms rarely target it directly.

To the extent that one would criticize *Ethics*, there are several possible motivations. The most direct would be to tackle the work's theological stance, which goes against anyone who would argue on behalf of the separation of God from the world, or on behalf of a purpose to the universe, since the book attacks both those ideas.[9] Another critique could be made based on a pragmatist* approach, which seeks to justify philosophical claims by observing the practical results of their application. Spinoza's system would appear to clash with pragmatism, since his approach is completely self-referential and its eternal laws allow no gaps or possibilities for change concerning either nature or humanity.

---

1. Benedict XVI, *Caritas in Veritate*, sec. 34, in Angela C. Miceli, "Alternative Foundations: A Dialogue with Modernity and the Papacy of Benedict XVI," *Perspectives on Political Science* 41, no. 1 (2012): 27.

2. Joshua J. McElwee, "Pope Francis: I would love a church that is poor," *National Catholic Reporter*, March 16, 2013, accessed on May 30, 2013, http://ncronline.org/blogs/pope-francis-i-would-love-church-poor.

3. McElwee, "Pope Francis."

4. See: Miceli, "Alternative Foundations."

5. Leonardo Boff, "Pope Benedict XVI is Leading the Church Astray," *International Press Service*, September 13, 2007, accessed on September 26, 2013, http://www.ipsnews.net/2007/09/pope-benedict-xvi-is-leading-the-church-astray/.

6. Simon C. Kim, "Theology of Context as the Theological Method of Virgilio Elizondo and Gustavo Gutiérrez," PhD diss., Catholic University of America, 2011, 300–8.

7. See: Paul E. Sigmund, *Liberation Theology at the Crossroads: Democracy or Revolution?* (New York: Oxford University Press, 1990); and Frederick Sontag, "Political Violence and Liberation Theology," *Journal of the Evangelical Theological Society* 33, no. 1 (March 1990): 85–94.

8. Joseph A. Varacalli, "A Catholic Sociological Critique of Gustavo Gutiérrez's *A Theology of Liberation*: A Review Essay," *The Catholic Social Science Review* 1 (1996): 175.

9. Kim, "Theology of Context," 295.

MODULE 12
WHERE NEXT?

KEY POINTS

* Spinoza's text remains the focus of intense attention, particularly in its application to two areas: scientific and religious naturalism,* and social and political theory.

* Insights from *Ethics* have been applied to widely varied fields, including political theory, psychoanalysis, cognitive science, and ethology* (an investigation into human and social behavior from a biological outlook).

* *Ethics* is a key text in the history of philosophy due to its explorations of pantheism,* monism,* and determinism,* as well as through its particular answer to the question of how to live a good life.

Potential

Benedictus de Spinoza's *Ethics* continues to attract intense attention, and there is little reason to doubt that such interest will fall off any time soon. How interest translates into influence is a different question, however. The answer depends on the parts of the work being analyzed, as well as the projects and purposes to which a given commentary contributes. As to where things go from here, two particularly compelling areas stand out. The first concerns explorations in scientific and religious naturalism and the second concerns social and political theory. For example, Spinoza's naturalism has been employed as a resource for contemporary philosophers of science,[1] some of whom admire its exclusion of God as an external agent acting on the world. At the same time, philosophers of religion, particularly Robert Corrington,* argue that Spinoza's naturalism opens nature up to theological enquiry.[2] Regarding politics, the anti-hierarchical tone of *Ethics*, in both form and content, as well as its resistance to teleological* thinking of all kinds, has proved favorable to the efforts of various contemporary French thinkers to radically critique existing ways of thinking about social institutions and language.[3]

> *"Obscure though Spinoza's ideas may be, there is no doubt that he was deeply committed to elucidating our everyday experience. Spinoza's metaphysics and epistemology make way for a kind of anthropology: a philosophy of human nature and a theory of how human beings relate to one another. Spinoza gives us tools for understanding ourselves and strategies for living well, something that few philosophers since the Greeks have attempted to provide."*
> —— Beth Lord, "Introduction," *Spinoza Beyond Philosophy*

Future Directions

There have been various twists and turns in the history of the reception of *Ethics*, from its rejection as atheistic* to its reevaluation as a key philosophical work through the pantheism controversy.* Its insights relate to many fields, from metaphysics* and epistemology* to political theory, psychoanalysis, ethology, and more. In its various forms, Spinoza's influence has been unpredictable, and in this sense, an open mind is needed when talking of future directions in the study of *Ethics*.

One example of a recent and relatively unexpected use to which Spinoza's thought has been put is found in the work of the neuroscientist Antonio Damasio.* His research, in contrast to the mainstream in cognitive science (the study of the mind), finds a neuroscientific explanation for affects (or emotions). In a series of books that are to an extent popularizations of more rigorous scientific work, Damasio rejects the Cartesian* focus on cognition (thinking) in recent neuroscientific work,[4] advancing instead a Spinozist focus on affect.[5] Drawing directly from Spinoza's psychology, he undertakes a neurology of joy and sorrow (that is, examining the neurological basis of those emotions). Though Damasio's particular views have been subject to criticism and rejection from various sides (including philosophers),[6] the general project of a neuroscientific analysis of affect is one that will continue to flourish. Such a general project also relates to technological advances, for instance, the field of affective computing, which studies the simulation of affects by computers. Damasio may not have got his Spinoza altogether right,[7] but he has taken one step in rightly implicating him in debates about these matters.

Summary

As long as philosophers continue to argue about mind, being, knowledge, or ethics, Benedictus de Spinoza will be read. His arguments on the perennial problems of philosophy are certainly original, and have influenced enough scholars of equal fame to Spinoza himself to ensure that *Ethics* appears in future philosophy texts. The fact that vital movements in contemporary philosophy incorporate close readings of *Ethics* indicates that even the work's complex details are likely to be the subject of ongoing study.

Spinoza's life is also worthy of continued attention. At a time in which one could be imprisoned or killed for holding unorthodox views on politics and religion, he refused to compromise. He saw it as his duty to seek the truth and investigate nature and the achievement of human happiness. Having been dramatically expelled from the Marrano* community of his youth, he lived a relatively isolated life as a glass lens-grinder in the Dutch countryside rather than give up his radical views. Many of those who knew him personally—even if they disagreed with his opinions—testified to his even temperament and congenial personality.

Unique in its form, breadth, and the radicalism of its content, Spinoza's work takes a varied mix of classical, medieval, and contemporary influences and weaves them into an intact philosophical system complete with its own methodology. In it, Spinoza asks how we maximize well-being in a deterministic universe without free will, and with no pride of place for humanity. The claims of *Ethics* cannot be separated from the form in which they are presented, which is a complex and self-referential framework of definitions and propositions. Spinoza is seeking nothing less than a new way of looking at minds, bodies, nature, and God, transforming them all into a single substance that stands as one of the most radical approaches to be found in the Western tradition. Spinoza's God, rather than being an entity full of emotions, goals, plans, and free will to act on the world, has none of these; He exists, rather, as a single substance uniting the physical and mental realms and identical with the eternal laws that govern all things indifferently, human or otherwise.

1. See: Marjorie G. Grene and Debra Nails (eds.), *Spinoza and the Sciences* (Dordrecht: Kluwer, 1986).

2. Robert S. Corrington, *Ecstatic Naturalism: Signs of the World* (Indianapolis: Indiana University Press, 1994).

3. See: Louis Althusser, *For Marx*, trans. Ben Brewster, (CITY: Verso, 1969), 78; Gilles Deleuze, *Spinoza: Practical Philosophy*, trans. Robert Hurley (San Francisco: City Lights Books, 1988), 69–70.

4. Antonio R. Damasio, *Descartes' Error: Emotion, Reason, and the Human Brain* (New York: Harper Perennial, 1995).

5. Antonio R. Damasio, *Looking for Spinoza: Joy, Sorrow, and the Feeling Brain* (Orlando, FL: Harvest, 2003).

6. See: M. R. Bennett, and P. M. S. Hacker, *Philosophical Foundations of Neuroscience* (Malden, MA: Wiley-Blackwell, 2003).

7. Ian Hacking, "Minding the Brain," *The New York Review of Books*, June 24, 2004, http://www.nybooks.com/articles/archives/2004/jun/24/minding-the-brain/.

 **GLOSSARY OF TERMS**

1. **Altruism:** a concern for the well-being of others.

2. **Anthropomorphism:** attributing human characteristics to a non-human being (for instance, imagining that a car's lights are its eyes, or picturing a god as a large person).

3. **Atheism:** the denial of the existence of God.

4. **Axiom:** a self-evident assumption on which other deductions can be based.

5. **Cartesian:** derived from the Latinized version of the name of René Descartes (Cartesius), an adjective used to indicate ideas influenced by Descartes's work.

6. **Causation:** in philosophy, the study of the relationship between cause and effect.

7. ***Cherem***: a Hebrew term for excommunication, a shunning or expulsion from the community of the faithful.

8. ***Conatus***: in Spinoza's work, a term that stands for the force by which each thing strives to persevere in its being.

9. **Determinism:** the philosophical view that all events are determined by causes that could not have produced any other action. This involves the denial of the existence of free will, since determinists hold that human action is determined in the same way as all other action.

10. **Empiricism:** a strand of philosophy that emphasizes sensory experience, such as seeing or hearing, as the primary—perhaps the only—source of knowledge. It also refers to knowledge gained through observation and experimentation.

11. **Enlightenment:** an intellectual movement, primarily during the eighteenth century, that championed reason, challenged traditional political authority, and called for the spreading of knowledge among the population. Spinoza is often seen as an early promoter of radical enlightenment.

12. **Epistemology:** an area of philosophy that addresses questions relating to the nature and limits of knowledge.

13. **Ethology:** the study of human behavior and social organization from a biological perspective.

14. **Excommunication:** a shunning or expulsion from the community of the faithful.

15. **Final cause:** a type of cause discussed by Aristotle. He distinguishes, among other types of causes, between efficient and final causes. The notion of efficient cause is the current common-sense notion of cause: A is the cause of B if in some sense A preceded B in time, and B was its product or effect. Final causes (Greek, *teloi*), on the other hand, refer to what a thing is for, its purpose. For example, a boat's efficient cause might be the boat-maker who made it, while its final cause might be to sail in the sea.

16. **Geometrical method:** in Latin, *geometrico* (translated as "geometrical manner"), a particular method of proof found originally in Euclid's geometry, and applied by Spinoza to the writing of philosophy. It begins by offering definitions and axioms, and proceeds by developing proofs and demonstrating them.

17. **German idealism:** refers to a tradition in German philosophy that followed in the wake of Immanuel Kant's *Critique of Pure Reason*. Broadly put, thinkers within the tradition, which included such figures as G. W. F. Hegel and Friedrich Schelling, sought to investigate the conditions through which the mind interacts with the world.

18. **Heresy:** a deviation from the accepted theological views of a particular religion.

19. **Marranos:** Sephardic Jews in the Iberian Peninsula (Portugal and Spain) who had been ordered to convert to Christianity upon the defeat of the Muslim Moors in the region during the fifteenth century. Spinoza's family were Portuguese Marranos, formerly called "Espinosa," who had emigrated to Amsterdam in the 1590s.

20. **Mennonites:** a sect of Protestant Christianity, named after Menno Simons, that originated in the sixteenth century. Their rejection of infant baptism in

favor of the baptism of adult "believers" brought scorn and persecution from other Christian communities, Catholic as well as Protestant. Mennonites are known for their pacifist beliefs.

21. **Metaphysics:** a term derived from ancient Greek philosophy, particularly that of Aristotle; it is the branch of philosophy that investigates the nature of being.

22. **Monism:** the philosophical position that claims that basically only one entity, or thing, exists. Substance monism is the view that only one substance exists.

23. **Naturalism:** a range of philosophical perspectives sharing the basic belief that natural laws are the basis on which all phenomena may be explained.

24. *Natura naturans* **and** *natura naturata*: terms that translate from the original Latin respectively as "nature in the active sense" and "nature already created," or more accurately as "nature naturing" and "nature natured."This implies a relation of cause (*natura naturans*) and effect (*natura naturata*), or creator and creature.

25. **Ontology:** the part of philosophy that deals with questions regarding being— that is, what kinds of things, broadly speaking, are there, or exist, or what is reality made of?

26. **Panentheism:** the view that God is in everything that exists. It contrasts with pantheism in that pantheism holds that God *is* everything in the universe (God *is* nature), while panentheism holds that God is *in* everything in the universe (God is *in* all nature).

27. **Pantheism:** the belief that everything in the universe is identical with the divine. Although the term does not itself appear in the *Ethics*, the work is considered one of philosophy's most prominent pantheistic texts.

28. **Pantheism controversy:** this began in the 1780s as a debate between Gotthold Lessing and Friedrich Heinrich Jacobi in which the former admitted embracing Spinoza's philosophy, as an alternative to the modes of thinking then dominant. The debate raised questions that not only shifted the way

in which Spinoza came to be viewed, but also marked a key transition in German thought from the Enlightenment of the eighteenth century to the Romanticism and idealism of the nineteenth.

29. **Pragmatism:** a philosophical tradition originating in the United States in the 1870s. Key figures include Charles S. Peirce, William James, and John Dewey, all of whom shared a commitment to observing and clarifying the effects of philosophical claims in lived experience.

30. **Rationalism:** a strand of Enlightenment thought that praised reason as both the means to access knowledge and indeed the repository of truth. Rationalism reached its height of influence in the late seventeenth and eighteenth centuries, and is contrasted with empiricism, knowledge gained through observation and experimentation.

31. **Reformation:** a process that began in 1517 when Martin Luther famously pinned his *Ninety-Five Theses* on the door of the church of Wittenberg, which would eventually lead to the Catholic Church's split from Protestantism.

32. **Renaissance:** a period in European history from the fourteenth until the seventeenth century, bridging the Middle Ages and modern history. It started in Italy as a cultural, intellectual, and artistic revival of ancient Greek culture. The word means literally "rebirth."

33. ***Res cogitans* and *res extensa*:** Descartes's Latin terms for "thinking thing" (*res cogitans*) and "extended thing" (*res extensa*), meant to indicate the essential attributes of mind, which is characterized by thought, and body, which is characterized by extension in space.

34. **Romanticism:** a philosophical, artistic, and literary movement that arose in Europe at the end of the eighteenth century and beginning of the nineteenth. A reaction against the rationalist emphases on universality, reason, and mathematics, Romanticism emphasized historical particularity, emotion, and the spiritual qualities of the natural world.

35. **Scholasticism:** a philosophical approach that dominated within medieval European universities and was characterized by logical disputations, as a

method of achieving the subtlest possible distinctions.

36. *Scholia*: the Greek word for "comments," used in the geometrical mode in which Spinoza writes to add less formal remarks.

37. *Scientia intuitiva*: a term that can be translated as "intellectual intuition," which, according to Spinoza, is a third kind of knowledge beyond that acquired through the senses or through pure reason.

38. **Stoicism:** a school of Greek philosophy founded by Zeno of Citium that praised control of one's passions as the highest good in life.

39. **Synod:** a church council or assembly, usually composed of senior members of the clergy.

40. **Teleology:** a term that derives from the Greek term *telos*, meaning "end" or "purpose," and which refers to both the study of purposes as well as any set of claims in which a purpose is implied as part of a thing or event's cause.

PEOPLE MENTIONED IN THE TEXT

1. **Aristotle (384–322 B.C.E.)** was a Greek philosopher who served as a pupil of Plato, founded the Lyceum in Athens, and composed treatises on almost every field of human knowledge, including ethics, aesthetics, metaphysics, and logic.

2. **Albert Balz (1887–1957)** was a contemporary American philosopher and historian of early modern philosophy.

3. **Pierre Bayle (1647–1706)** was a French philosopher who, following Descartes, advocated strict separation between faith and reason. His *Historical and Critical Dictionary*, first published in 1697, was widely read among European intellectuals at the turn of the eighteenth century.

4. **Jonathan Bennett (b. 1930)** is a contemporary British philosopher and historian of early modern philosophy.

5. **Jorge Luis Borges (1899–1986)** was an Argentinean writer and poet, one of the most important figures in twentieth-century literature.

6. **Robert S. Corrington (b. 1950)** is a contemporary American writer and philosopher.

7. **Edwin Curley (b. 1937)** is a contemporary American philosopher and Spinoza scholar.

8. **Antonio Damasio (b. 1944)** is a contemporary neuroscientist and author.

9. **Gilles Deleuze (1925–95)** was a contemporary French philosopher and historian of philosophy.

10. **René Descartes (1596–1650)** was a French mathematician and philosopher whose dualism between mind and world and whose philosophical method of clear and distinct ideas upon indubitable foundations marked the decisive turning point from medieval to modern philosophy, as well as a broader shift between philosophies of being toward those of consciousness.

11. **Albert Einstein (1879–1955)** was the most significant contributor to twentieth-century theoretical physics, known for developing the theory of relativity.

12. **Franciscus van den Enden (1602–74)** was most famous for having been Spinoza's teacher. He was, among other things, a political radical, poet, and philosopher. He spent a period of his life as a Jesuit, though he soon abandoned the order, and was thought by some to have been an atheist. He was convicted of having plotted to murder the French king Louis XIV and hanged outside the Bastille.

13. **Euclid of Alexandria (c. mid-fourth century–mid-third century B.C.E.)** was a Greek mathematician whose primary work, *Elements*, served as the textbook for math courses until the early twentieth century. He is considered the father of geometry.

14. **Sigmund Freud (1856–1939)** was an Austrian neurologist who is known as the father of psychoanalysis. In terms of intellectual history, Freud is significant, along with Karl Marx and Friedrich Nietzsche, for undermining the Enlightenment consensus that humans are fundamentally rational beings.

15. **Peter Gay (1923–2015)** was a contemporary American historian.

16. **Johann Wolfgang von Goethe (1749–1832)** was a German philosopher, writer, scientist, and statesman. He is known among other things for his association with the *Sturm und Drang* ("Storm and Drive") literary movement of the late eighteenth century.

17. **Felix Guattari (1930–92)** was a contemporary French psychotherapist, philosopher, and activist.

18. **Georg Wilhelm Friedrich Hegel (1770–1831)** was one of the most significant philosophers of the past two hundred years. Associated with the German idealist tradition initiated by Immanuel Kant, Hegel's brand of idealism, which he called absolute idealism, describes nature as an absolute system whose manifestations unfold across a logical historical path.

19. **Friedrich Heinrich Jacobi (1743–1819)** was a German post-Kantian philosopher and participant in the pantheism controversy.

20. **Gottfried Wilhelm von Leibniz (1646–1716)** was a German mathematician and philosopher who is credited, along with Isaac Newton, as an independent

co-inventor of calculus.

21. **Gotthold Lessing (1729–81)** was a German philosopher and poet whose work marked a high point in German Enlightenment thinking before its transition into Romanticism.

22. **John Locke (1632–1704)** was an English philosopher and political theorist whose writings are considered foundational to empiricist philosophy and modern liberalism.

23. **Maimonides (1135–1204)**, born Mosheh ben Maimon, was a twelfth-century philosopher and scholar of the Torah who lived in Spain and North Africa. Noted for his explications of Jewish law, Maimonides is also renowned for his synthesis of biblical and Aristotelian thought.

24. **Nicolas Malebranche (1638–1715)** was a Cartesian rationalist philosopher and priest. He is best known for his defense of a version of the doctrine known as occasionalism, which holds that mental events cannot cause bodily events, and vice versa; rather, each time the mind wills something, God causes bodily reality to make it so.

25. **Pierre-François Moreau (b. 1948)** is a contemporary French philosopher and historian of French philosophy.

26. **Friedrich Nietzsche (1844–1900)** was a German philosopher and cultural critic whose work radically challenged many of the most entrenched notions in Western thought, particularly regarding morality and religion.

27. **Bertrand Russell (1872–1970)** was a contemporary British philosopher and mathematician, and one of the twentieth century's leading public intellectuals.

28. **Friedrich Schelling (1775–1854)** was a German philosopher associated with the tradition of German idealism initiated by Immanuel Kant (1724–1804). He emphasized chance as an inescapable fact of nature.

29. **Jan de Witt (1625–72)** was a republican politician, first Statesman of the Dutch Republic from the 1650s until his death. He was forcibly deposed by royalists, who lynched him and his brother in 1672.

 WORKS CITED

1. Abadi, Marcelo. "Spinoza in Borges' Looking-Glass." *Studia Spinozana: An International and Interdisciplinary Series* 5 (1989): 29–42.

2. Alexander, Douglas. "Was Spinoza a Naturalist?" *Pacific Philosophical Quarterly* 96, no. 1 (2015): 77–99.

3. Althusser, Louis. *For Marx*. Translated by Ben Brewster. New York: Verso, 1969.

4. Balz, Albert G. A. *Idea and Essence in the Philosophies of Hobbes and Spinoza*. New York: Columbia University Press, 1918.

5. Bennett, Jonathan. *A Study of Spinoza's Ethics*. New York: Hackett, 1984.

6. Bennett, M. R., and P. M. S. Hacker. *Philosophical Foundations of Neuroscience*. Malden, MA: Wiley-Blackwell, 2003.

7. Bernard, Walter. "Freud and Spinoza." *Psychiatry* 9, no. 2 (1946): 99–108.

8. Buijs, M. E. J. "How to Make a Living God" (not yet published).

9. Copleston, F. C. "Pantheism in Spinoza and the German Idealists." *Philosophy* 21, no. 78 (1946): 42–56.

10. Corrington, Robert S. *Ecstatic Naturalism: Signs of the World*. Indianapolis: Indiana University Press, 1994.

11. Curley, Edwin. *Spinoza's Metaphysics: An Essay in Interpretation*. Cambridge, MA: Harvard University Press, 1969.

12. Damasio, Antonio. *Descartes' Error: Emotion, Reason, and the Human Brain*. New York: Harper Perennial, 1995.

13. ——. *Looking for Spinoza: Joy, Sorrow, and the Feeling Brain*. Orlando, FL: Harvest, 2003.

14. Davidson, Donald. "Spinoza's Causal Theory of the Affects." In *Desire and Affect: Spinoza as Psychologist*, ed. Yirmiyahu Yovel, 95–111. Leiden: Brill, 1999.

15. Deleuze, Gilles. *Expressionism in Philosophy: Spinoza*. Cambridge, MA: MIT Press, 1992.

16. ——. *Spinoza: Practical Philosophy*. Translated by Robert Hurley. San Francisco: City Lights Books, 1988.

17. Descartes, René. *Principles of Philosophy*. Translated by V. Rodger Miller and R. P. Miller. London: Reidel, 1983.

18. Dewey, John. "The Pantheism of Spinoza." *The Journal of Speculative Philosophy* 16, no.3 (1882): 249–257.

19. Fløistad, Guttorm. "Spinoza's Theory of Knowledge." *Inquiry* 12, nos.1–4 (1969): 41–65.

20. Gay, Peter. *The Enlightenment: A Comprehensive Anthology*. New York: Simon and Schuster, 1973.

21. Goetschel, Willi. *Spinoza's Modernity: Mendelssohn, Lessing, and Heine*. Madison, WI: University of Wisconsin Press, 2004.

22. Grene, Marjorie G., and Debra Nails, eds. *Spinoza and the Sciences*. Dordrecht: Kluwer, 1986.

23. Hacking, Ian. "Minding the Brain." *The New York Review of Books*, June 24, 2004. Accessed September 24, 2015. http://www.nybooks.com/articles/archives/2004/jun/24/minding-the-brain/.

24. Hampshire, Stuart. *Spinoza*. Manchester: Manchester University Press, 1956.

25. Hegel, G. W. F. *Lectures on the History of Philosophy, Volume 3: Medieval and Modern Philosophy*. Translated by E. S. Haldane and Frances H. Simson. Lincoln: University of Nebraska Press, 1995.

26. Jacobi, Friedrich Heinrich, and Gérard Vallée. *The Spinoza Conversations Between Lessing and Jacobi: Text with Excerpts from the Ensuing Controversy*. University Press of America, 1988.

27. James, Susan. "When Does Truth Matter? Spinoza on the Relation Between Theology and Philosophy." *European Journal of Philosophy* 20, no.1 (2012): 91–108.

28. Jarrett, Charles E. *Spinoza: A Guide for the Perplexed*. New York: Continuum, 2007.

29. Koistinen, Olli. "Causality, Intentionality, and Identity: Mind–Body Interaction." *Ratio* 9 no.1 (1996): 23–38.

30. Lloyd, Genevieve. *Routledge Philosophy GuideBook to Spinoza and the* Ethics. London: Routledge, 1996.

31. ——. "Spinoza's Environmental Ethics." *Inquiry* 23, no.3 (1980): 293–311.

32. Lord, B., ed. *Spinoza Beyond Philosophy*. Edinburgh: Edinburgh University Press, 2012.

33. Moreau, Pierre-François. *Spinoza: L'expérience et l'éternité*. Paris: Presses Universitaires de France, 1994.

34. Nietzsche, Friedrich Wilhelm. *Nietzsche: The Gay Science: With a Prelude in German Rhymes and an Appendix of Songs*. Translated by Josefine Nauckhoff. London: Cambridge University Press, 2001.

35. ——. *Selected Letters of Friedrich Nietzsche*. Edited and translated by Christopher Middleton. New York: Hackett Publishing, 1996.

36. Norris, Christofer. "Spinoza and the Conflict of Interpretations." In *Spinoza Now*, ed. Dimitris Vardoulakis, 3–38. Minnesota: University of Minnesota Press, 2011.

37. Peden, Knox. *Spinoza Contra Phenomenology: French Rationalism from Cavaillès to Deleuze*. California: Stanford University Press, 2014.

38. Ravven, H. M., and L. E. Goodman. *Jewish Themes in Spinoza's Philosophy*. New York: SUNY Press, 2012.

39. Russell, Bertrand. *History of Western Philosophy*. Oxon: Routledge, 2004.

40. Scruton, Roger. *Spinoza*. Oxford: Oxford University Press, 1986.

41. Spinoza, Baruch (Benedictus de). *The Correspondence of Spinoza*. New York, Russell & Russell, 1928.

42. ——. *Ethics*. In *A Spinoza Reader: The Ethics and Other Works*, translated by Edwin Curley. Princeton: Princeton University Press, 1996, 85–265.

43. ——. "Political Treatise." In *A Theologico-Political Treatise and a Political Treatise*, edited by Francesco Cordasco. New York: Courier Corporation, 2013.

44. ——. *The Principles of Cartesian Philosophy: With Metaphysical Thoughts*. Edited by Samuel Shirley. New York, Hackett Publishing, 1998.

45. ——.*Theological-Political Treatise*. Cambridge, Cambridge University Press, 2007.

46. ——. "The Treatise on the Emendation of the Intellect." In *Ethics: With the Treatise on the Emendation of the Intellect and Selected Letters*, translated by Samuel Shirley. Indianapolis: Hackett Publishing, 1992.

47. Westphal, Merold. "Hegel between Spinoza and Derrida." In *Hegel's History of Philosophy: New Interpretations*, edited by David Duquette. Albany, NY: State

of New York Press, 2003.

48. Wolf, Abraham. "Spinoza's Conception of Attributes." In *Studies in Spinoza, Critical and Interpretive Essays*, edited by S. Paul Kashap. California: University of California Press, 1972.

原书作者简介

巴鲁赫·斯宾诺莎 1632 年出生于阿姆斯特丹的马拉诺犹太社区。马拉诺人是为了逃避葡萄牙的迫害而移居荷兰的。1656 年，这个社区因"可怕的异端邪说"——明显是针对斯宾诺莎非正统的上帝观和宗教观——将他革出教门。斯宾诺莎后来隐居在一个荷兰小镇，当起了磨镜匠。但他没有放弃写作，只不过把大部分作品的署名改为他希伯来语名字的拉丁语版——本尼迪克特·德·斯宾诺莎。他被犹太教和基督教当局看作无神论者和异端，遭到各界谴责。但一个世纪后，学者们开始注意到他的作品，他也逐渐被视为最独特、最重要的西方哲学家之一。1677 年斯宾诺莎去世，年仅 44 岁。

本书作者简介

盖里·斯雷特博士，牛津大学神学和宗教学博士，现执教于得克萨斯州的圣爱德华大学。

安德里亚·弗拉希米斯，著有《分析哲学与大陆哲学的相遇》（2013），现执教于塞浦路斯大学。

世界名著中的批判性思维

《世界思想宝库钥匙丛书》致力于深入浅出地阐释全世界著名思想家的观点，不论是谁、在何处都能了解到，从而推进批判性思维发展。

《世界思想宝库钥匙丛书》与世界顶尖大学的一流学者合作，为一系列学科中最有影响的著作推出新的分析文本，介绍其观点和影响。在这一不断扩展的系列中，每种选入的著作都代表了历经时间考验的思想典范。通过为这些著作提供必要背景、揭示原作者的学术渊源以及说明这些著作所产生的影响，本系列图书希望让读者以新视角看待这些划时代的经典之作。读者应学会思考、运用并挑战这些著作中的观点，而不是简单接受它们。

ABOUT THE AUTHOR OF THE ORIGINAL WORK

Baruch Spinoza was born in 1632 into Amsterdam's Marrano community of Jews who had fled persecution in Portugal. In 1656, the community excommunicated him for "horrible heresies"—apparently due to his unorthodox views on God and religion. Spinoza retreated to a small Dutch town where he worked as a glass lens-grinder. However, he continued writing, signing most of his works as Benedictus, the Latin version of his Hebrew first name. Spinoza was widely denounced as an atheist and heretic by both Jewish and Christian religious authorities, but a century later scholars began to take note of his work and he gradually came to be seen as one of the most original and important of all Western philosophers. Spinoza died in 1677, aged just 44.

ABOUT THE AUTHORS OF THE ANALYSIS

Dr Gary Slater holds a DPhil in theology and religion from the University of Oxford. He is currently a faculty member at St Edwards University, Texas.
Andreas Vrahimis is the author of *Encounters between Analytic and Continental Philosophy* (2013). He teaches at the University of Cyprus.

ABOUT MACAT
GREAT WORKS FOR CRITICAL THINKING

Macat is focused on making the ideas of the world's great thinkers accessible and comprehensible to everybody, everywhere, in ways that promote the development of enhanced critical thinking skills.

It works with leading academics from the world's top universities to produce new analyses that focus on the ideas and the impact of the most influential works ever written across a wide variety of academic disciplines. Each of the works that sit at the heart of its growing library is an enduring example of great thinking. But by setting them in context—and looking at the influences that shaped their authors, as well as the responses they provoked—Macat encourages readers to look at these classics and game-changers with fresh eyes. Readers learn to think, engage and challenge their ideas, rather than simply accepting them.

批判性思维与《伦理学》

首要批判性思维技巧：理性化思维

次要批判性思维技巧：阐释

      巴鲁赫·斯宾诺莎的《伦理学》是一本充满长篇论辩推理的杰作。在上帝的本质、宇宙、自由意志以及人的道德等话题上，《伦理学》的论辩毫不妥协、直截了当，因此它具有广泛的影响力，在西方哲学著作中占有重要的一席之地。

      虽然《伦理学》是西方哲学经典著作中最厚重、最难懂的著作之一，但斯宾诺莎组织、构建论述的独特方法颇为知名。正如其全名——《用几何学方法作论证的伦理学》——所示，斯宾诺莎用严谨的数学式命题的格式来展开论述，就像古希腊数学家欧几里得用几何学命题来展开基本几何原理那样。

      斯宾诺莎的这部杰作论述方法自成体系，具备了高超推理技巧的关键要素。推理的关键用处在于使论述组织严密，结论有说服力，推理过程逻辑清晰。就像一位数学家展开一段几何学证明那样，斯宾诺莎要展示的是一种全面的人类生存哲学。这种努力也影响了一代又一代的后辈哲学家们。

CRITICAL THINKING AND *ETHICS*

• Primary critical thinking skill: REASONING

• Secondary critical thinking skill: INTERPRETATION

Baruch Spinoza's *Ethics* is a dense masterpiece of sustained argumentative reasoning. It earned its place as one of the most important and influential books in Western philosophy by virtue of its uncompromisingly direct arguments about the nature of God, the universe, free will, and human morals.

Though it remains one of the densest and most challenging texts in the entire canon of Western philosophy, *Ethics* is also famous for Spinoza's unique approach to ordering and constructing its arguments. As its full title—*Ethics, Demonstrated in Geometrical Order*—suggests, Spinoza decided to use the rigorous format of mathematical-style propositions to lay out his arguments, just as the Ancient Greek mathematician Euclid had used geometrical propositions to lay out the basic rules of geometry.

In choosing such a systematic method, Spinoza's masterwork shows the crucial aspects of good reasoning skills being employed at the highest level. The key use of reasoning is the production of an argument that is well-organised, supports its conclusions and proceeds logically towards its end. Just as a mathematician might demonstrate a geometrical proof, Spinoza sought to lay out a comprehensive philosophy for human existence—an attempt that has influenced generations of philosophers since.

《世界思想宝库钥匙丛书》简介

《世界思想宝库钥匙丛书》致力于为一系列在各领域产生重大影响的人文社科类经典著作提供独特的学术探讨。每一本读物都不仅仅是原经典著作的内容摘要，而是介绍并深入研究原经典著作的学术渊源、主要观点和历史影响。这一丛书的目的是提供一套学习资料，以促进读者掌握批判性思维，从而更全面、深刻地去理解重要思想。

每一本读物分为 3 个部分：学术渊源、学术思想和学术影响，每个部分下有 4 个小节。这些章节旨在从各个方面研究原经典著作及其反响。

由于独特的体例，每一本读物不但易于阅读，而且另有一项优点：所有读物的编排体例相同，读者在进行某个知识层面的调查或研究时可交叉参阅多本该丛书中的相关读物，从而开启跨领域研究的路径。

为了方便阅读，每本读物最后还列出了术语表和人名表（在书中则以星号 * 标记），此外还有参考文献。

《世界思想宝库钥匙丛书》与剑桥大学合作，理清了批判性思维的要点，即如何通过 6 种技能来进行有效思考。其中 3 种技能让我们能够理解问题，另 3 种技能让我们有能力解决问题。这 6 种技能合称为"批判性思维 PACIER 模式"，它们是：

分析：了解如何建立一个观点；

评估：研究一个观点的优点和缺点；

阐释：对意义所产生的问题加以理解；

创造性思维：提出新的见解，发现新的联系；

解决问题：提出切实有效的解决办法；

理性化思维：创建有说服力的观点。

THE MACAT LIBRARY

The Macat Library is a series of unique academic explorations of seminal works in the humanities and social sciences — books and papers that have had a significant and widely recognised impact on their disciplines. It has been created to serve as much more than just a summary of what lies between the covers of a great book. It illuminates and explores the influences on, ideas of, and impact of that book. Our goal is to offer a learning resource that encourages critical thinking and fosters a better, deeper understanding of important ideas.

Each publication is divided into three Sections: Influences, Ideas, and Impact. Each Section has four Modules. These explore every important facet of the work, and the responses to it.

This Section-Module structure makes a Macat Library book easy to use, but it has another important feature. Because each Macat book is written to the same format, it is possible (and encouraged!) to cross-reference multiple Macat books along the same lines of inquiry or research. This allows the reader to open up interesting interdisciplinary pathways.

To further aid your reading, lists of glossary terms and people mentioned are included at the end of this book (these are indicated by an asterisk [*] throughout) — as well as a list of works cited.

Macat has worked with the University of Cambridge to identify the elements of critical thinking and understand the ways in which six different skills combine to enable effective thinking.

Three allow us to fully understand a problem; three more give us the tools to solve it. Together, these six skills make up the PACIER model of critical thinking. They are:

ANALYSIS — understanding how an argument is built
EVALUATION — exploring the strengths and weaknesses of an argument
INTERPRETATION — understanding issues of meaning
CREATIVE THINKING — coming up with new ideas and fresh connections
PROBLEM-SOLVING — producing strong solutions
REASONING — creating strong arguments

"《世界思想宝库钥匙丛书》提供了独一无二的跨学科学习和研究工具。它介绍那些革新了各自学科研究的经典著作，还邀请全世界一流专家和教育机构进行严谨的分析，为每位读者打开世界顶级教育的大门。"

—— 安德烈亚斯·施莱歇尔，
经济合作与发展组织教育与技能司司长

"《世界思想宝库钥匙丛书》直面大学教育的巨大挑战……他们组建了一支精干而活跃的学者队伍，来推出在研究广度上颇具新意的教学材料。"

—— 布罗尔斯教授、勋爵，剑桥大学前校长

"《世界思想宝库钥匙丛书》的愿景令人赞叹。它通过分析和阐释那些曾深刻影响人类思想以及社会、经济发展的经典文本，提供了新的学习方法。它推动批判性思维，这对于任何社会和经济体来说都是至关重要的。这就是未来的学习方法。"

—— 查尔斯·克拉克阁下，英国前教育大臣

"对于那些影响了各自领域的著作，《世界思想宝库钥匙丛书》能让人们立即了解到围绕那些著作展开的评论性言论，这让该系列图书成为在这些领域从事研究的师生们不可或缺的资源。"

—— 威廉·特朗佐教授，加利福尼亚大学圣地亚哥分校

"Macat offers an amazing first-of-its-kind tool for interdisciplinary learning and research. Its focus on works that transformed their disciplines and its rigorous approach, drawing on the world's leading experts and educational institutions, opens up a world-class education to anyone."

—— Andreas Schleicher, Director for Education and Skills, Organisation for Economic Co-operation and Development

"Macat is taking on some of the major challenges in university education... They have drawn together a strong team of active academics who are producing teaching materials that are novel in the breadth of their approach."

—— Prof Lord Broers, former Vice-Chancellor of the University of Cambridge

"The Macat vision is exceptionally exciting. It focuses upon new modes of learning which analyse and explain seminal texts which have profoundly influenced world thinking and so social and economic development. It promotes the kind of critical thinking which is essential for any society and economy. This is the learning of the future."

—— Rt Hon Charles Clarke, former UK Secretary of State for Education

"The Macat analyses provide immediate access to the critical conversation surrounding the books that have shaped their respective discipline, which will make them an invaluable resource to all of those, students and teachers, working in the field."

—— Prof William Tronzo, University of California at San Diego

⚷ The Macat Library
世界思想宝库钥匙丛书

TITLE	中文书名	类别
An Analysis of Arjun Appadurai's *Modernity at Large: Cultural Dimensions of Globalization*	解析阿尔君·阿帕杜莱《消失的现代性：全球化的文化维度》	人类学
An Analysis of Claude Lévi-Strauss's *Structural Anthropology*	解析克劳德·列维-斯特劳斯《结构人类学》	人类学
An Analysis of Marcel Mauss's *The Gift*	解析马塞尔·莫斯《礼物》	人类学
An Analysis of Jared M. Diamond's *Guns, Germs, and Steel: The Fate of Human Societies*	解析贾雷德·M.戴蒙德《枪炮、病菌与钢铁：人类社会的命运》	人类学
An Analysis of Clifford Geertz's *The Interpretation of Cultures*	解析克利福德·格尔茨《文化的解释》	人类学
An Analysis of Philippe Ariès's *Centuries of Childhood: A Social History of Family Life*	解析菲力浦·阿利埃斯《儿童的世纪：旧制度下的儿童和家庭生活》	人类学
An Analysis of W. Chan Kim & Renée Mauborgne's *Blue Ocean Strategy*	解析金伟灿/勒妮·莫博涅《蓝海战略》	商业
An Analysis of John P. Kotter's *Leading Change*	解析约翰·P.科特《领导变革》	商业
An Analysis of Michael E. Porter's *Competitive Strategy: Techniques for Analyzing Industries and Competitors*	解析迈克尔·E.波特《竞争战略：分析产业和竞争对手的技术》	商业
An Analysis of Jean Lave & Etienne Wenger's *Situated Learning: Legitimate Peripheral Participation*	解析琼·莱夫/艾蒂纳·温格《情境学习：合法的边缘性参与》	商业
An Analysis of Douglas McGregor's *The Human Side of Enterprise*	解析道格拉斯·麦格雷戈《企业的人性面》	商业
An Analysis of Milton Friedman's *Capitalism and Freedom*	解析米尔顿·弗里德曼《资本主义与自由》	商业
An Analysis of Ludwig von Mises's *The Theory of Money and Credit*	解析路德维希·冯·米塞斯《货币和信用理论》	经济学
An Analysis of Adam Smith's *The Wealth of Nations*	解析亚当·斯密《国富论》	经济学
An Analysis of Thomas Piketty's *Capital in the Twenty-First Century*	解析托马斯·皮凯蒂《21世纪资本论》	经济学
An Analysis of Nassim Nicholas Taleb's *The Black Swan: The Impact of the Highly Improbable*	解析纳西姆·尼古拉斯·塔勒布《黑天鹅：如何应对不可预知的未来》	经济学
An Analysis of Ha-Joon Chang's *Kicking Away the Ladder*	解析张夏准《富国陷阱：发达国家为何踢开梯子》	经济学
An Analysis of Thomas Robert Malthus's *An Essay on the Principle of Population*	解析托马斯·罗伯特·马尔萨斯《人口论》	经济学

An Analysis of John Maynard Keynes's *The General Theory of Employment, Interest and Money*	解析约翰·梅纳德·凯恩斯《就业、利息和货币通论》	经济学
An Analysis of Milton Friedman's *The Role of Monetary Policy*	解析米尔顿·弗里德曼《货币政策的作用》	经济学
An Analysis of Burton G. Malkiel's *A Random Walk Down Wall Street*	解析伯顿·G.马尔基尔《漫步华尔街》	经济学
An Analysis of Friedrich A. Hayek's *The Road to Serfdom*	解析弗里德里希·A.哈耶克《通往奴役之路》	经济学
An Analysis of Charles P. Kindleberger's *Manias, Panics, and Crashes: A History of Financial Crises*	解析查尔斯·P.金德尔伯格《疯狂、惊恐和崩溃：金融危机史》	经济学
An Analysis of Amartya Sen's *Development as Freedom*	解析阿马蒂亚·森《以自由看待发展》	经济学
An Analysis of Rachel Carson's *Silent Spring*	解析蕾切尔·卡森《寂静的春天》	地理学
An Analysis of Charles Darwin's *On the Origin of Species: by Means of Natural Selection, or The Preservation of Favoured Races in the Struggle for Life*	解析查尔斯·达尔文《物种起源》	地理学
An Analysis of World Commission on Environment and Development's *The Brundtland Report: Our Common Future*	解析世界环境与发展委员会《布伦特兰报告：我们共同的未来》	地理学
An Analysis of James E. Lovelock's *Gaia: A New Look at Life on Earth*	解析詹姆斯·E.拉伍洛克《盖娅：地球生命的新视野》	地理学
An Analysis of Paul Kennedy's *The Rise and Fall of the Great Powers: Economic Change and Military Conflict from 1500–2000*	解析保罗·肯尼迪《大国的兴衰：1500—2000年的经济变革与军事冲突》	历史
An Analysis of Janet L. Abu-Lughod's *Before European Hegemony: The World System A. D. 1250–1350*	解析珍妮特·L.阿布–卢格霍德《欧洲霸权之前：1250—1350年的世界体系》	历史
An Analysis of Alfred W. Crosby's *The Columbian Exchange: Biological and Cultural Consequences of 1492*	解析艾尔弗雷德·W.克罗斯比《哥伦布大交换：1492年以后的生物影响和文化冲击》	历史
An Analysis of Tony Judt's *Postwar: A History of Europe since 1945*	解析托尼·朱特《战后欧洲史》	历史
An Analysis of Richard J. Evans's *In Defence of History*	解析理查德·J.艾文斯《捍卫历史》	历史
An Analysis of Eric Hobsbawm's *The Age of Revolution: Europe 1789–1848*	解析艾瑞克·霍布斯鲍姆《革命的年代：欧洲1789—1848年》	历史

An Analysis of Roland Barthes's *Mythologies*	解析罗兰·巴特《神话学》	文学与批判理论
An Analysis of Simone de Beauvoir's *The Second Sex*	解析西蒙娜·德·波伏娃《第二性》	文学与批判理论
An Analysis of Edward W. Said's *Orientalism*	解析爱德华·W.萨义德《东方主义》	文学与批判理论
An Analysis of Virginia Woolf's *A Room of One's Own*	解析弗吉尼亚·伍尔芙《一间自己的房间》	文学与批判理论
An Analysis of Judith Butler's *Gender Trouble*	解析朱迪斯·巴特勒《性别麻烦》	文学与批判理论
An Analysis of Ferdinand de Saussure's *Course in General Linguistics*	解析费尔迪南·德·索绪尔《普通语言学教程》	文学与批判理论
An Analysis of Susan Sontag's *On Photography*	解析苏珊·桑塔格《论摄影》	文学与批判理论
An Analysis of Walter Benjamin's *The Work of Art in the Age of Mechanical Reproduction*	解析瓦尔特·本雅明《机械复制时代的艺术作品》	文学与批判理论
An Analysis of W. E. B. Du Bois's *The Souls of Black Folk*	解析W.E.B.杜波依斯《黑人的灵魂》	文学与批判理论
An Analysis of Plato's *The Republic*	解析柏拉图《理想国》	哲学
An Analysis of Plato's *Symposium*	解析柏拉图《会饮篇》	哲学
An Analysis of Aristotle's *Metaphysics*	解析亚里士多德《形而上学》	哲学
An Analysis of Aristotle's *Nicomachean Ethics*	解析亚里士多德《尼各马可伦理学》	哲学
An Analysis of Immanuel Kant's *Critique of Pure Reason*	解析伊曼努尔·康德《纯粹理性批判》	哲学
An Analysis of Ludwig Wittgenstein's *Philosophical Investigations*	解析路德维希·维特根斯坦《哲学研究》	哲学
An Analysis of G. W. F. Hegel's *Phenomenology of Spirit*	解析G.W.F.黑格尔《精神现象学》	哲学
An Analysis of Baruch Spinoza's *Ethics*	解析巴鲁赫·斯宾诺莎《伦理学》	哲学
An Analysis of Hannah Arendt's *The Human Condition*	解析汉娜·阿伦特《人的境况》	哲学
An Analysis of G. E. M. Anscombe's *Modern Moral Philosophy*	解析G.E.M.安斯康姆《现代道德哲学》	哲学
An Analysis of David Hume's *An Enquiry Concerning Human Understanding*	解析大卫·休谟《人类理解研究》	哲学

An Analysis of Søren Kierkegaard's *Fear and Trembling*	解析索伦·克尔凯郭尔《恐惧与战栗》	哲学
An Analysis of René Descartes's *Meditations on First Philosophy*	解析勒内·笛卡尔《第一哲学沉思录》	哲学
An Analysis of Friedrich Nietzsche's *On the Genealogy of Morality*	解析弗里德里希·尼采《论道德的谱系》	哲学
An Analysis of Gilbert Ryle's *The Concept of Mind*	解析吉尔伯特·赖尔《心的概念》	哲学
An Analysis of Thomas Kuhn's *The Structure of Scientific Revolutions*	解析托马斯·库恩《科学革命的结构》	哲学
An Analysis of John Stuart Mill's *Utilitarianism*	解析约翰·斯图亚特·穆勒《功利主义》	哲学
An Analysis of Aristotle's *Politics*	解析亚里士多德《政治学》	政治学
An Analysis of Niccolò Machiavelli's *The Prince*	解析尼科洛·马基雅维利《君主论》	政治学
An Analysis of Karl Marx's *Capital*	解析卡尔·马克思《资本论》	政治学
An Analysis of Benedict Anderson's *Imagined Communities*	解析本尼迪克特·安德森《想象的共同体》	政治学
An Analysis of Samuel P. Huntington's *The Clash of Civilizations and the Remaking of World Order*	解析塞缪尔·P.亨廷顿《文明的冲突与世界秩序的重建》	政治学
An Analysis of Alexis de Tocqueville's *Democracy in America*	解析阿列克西·德·托克维尔《论美国的民主》	政治学
An Analysis of John A. Hobson's *Imperialism: A Study*	解析约翰·A.霍布森《帝国主义》	政治学
An Analysis of Thomas Paine's *Common Sense*	解析托马斯·潘恩《常识》	政治学
An Analysis of John Rawls's *A Theory of Justice*	解析约翰·罗尔斯《正义论》	政治学
An Analysis of Francis Fukuyama's *The End of History and the Last Man*	解析弗朗西斯·福山《历史的终结与最后的人》	政治学
An Analysis of John Locke's *Two Treatises of Government*	解析约翰·洛克《政府论》	政治学
An Analysis of Sun Tzu's *The Art of War*	解析孙武《孙子兵法》	政治学
An Analysis of Henry Kissinger's *World Order: Reflections on the Character of Nations and the Course of History*	解析亨利·基辛格《世界秩序》	政治学
An Analysis of Jean-Jacques Rousseau's *The Social Contract*	解析让-雅克·卢梭《社会契约论》	政治学

An Analysis of Odd Arne Westad's *The Global Cold War: Third World Interventions and the Making of Our Times*	解析文安立《全球冷战：美苏对第三世界的干涉与当代世界的形成》	政治学
An Analysis of Sigmund Freud's *The Interpretation of Dreams*	解析西格蒙德·弗洛伊德《梦的解析》	心理学
An Analysis of William James' *The Principles of Psychology*	解析威廉·詹姆斯《心理学原理》	心理学
An Analysis of Philip Zimbardo's *The Lucifer Effect*	解析菲利普·津巴多《路西法效应》	心理学
An Analysis of Leon Festinger's *A Theory of Cognitive Dissonance*	解析利昂·费斯汀格《认知失调论》	心理学
An Analysis of Richard H. Thaler & Cass R. Sunstein's *Nudge: Improving Decisions about Health, Wealth, and Happiness*	解析理查德·H.泰勒/卡斯·R.桑斯坦《助推：如何做出有关健康、财富和幸福的更优决策》	心理学
An Analysis of Gordon Allport's *The Nature of Prejudice*	解析高尔登·奥尔波特《偏见的本质》	心理学
An Analysis of Steven Pinker's *The Better Angels of Our Nature: Why Violence Has Declined*	解析斯蒂芬·平克《人性中的善良天使：暴力为什么会减少》	心理学
An Analysis of Stanley Milgram's *Obedience to Authority*	解析斯坦利·米尔格拉姆《对权威的服从》	心理学
An Analysis of Betty Friedan's *The Feminine Mystique*	解析贝蒂·弗里丹《女性的奥秘》	心理学
An Analysis of David Riesman's *The Lonely Crowd: A Study of the Changing American Character*	解析大卫·理斯曼《孤独的人群：美国人社会性格演变之研究》	社会学
An Analysis of Franz Boas's *Race, Language and Culture*	解析弗朗兹·博厄斯《种族、语言与文化》	社会学
An Analysis of Pierre Bourdieu's *Outline of a Theory of Practice*	解析皮埃尔·布尔迪厄《实践理论大纲》	社会学
An Analysis of Max Weber's *The Protestant Ethic and the Spirit of Capitalism*	解析马克斯·韦伯《新教伦理与资本主义精神》	社会学
An Analysis of Jane Jacobs's *The Death and Life of Great American Cities*	解析简·雅各布斯《美国大城市的死与生》	社会学
An Analysis of C. Wright Mills's *The Sociological Imagination*	解析C.赖特·米尔斯《社会学的想象力》	社会学
An Analysis of Robert E. Lucas Jr.'s *Why Doesn't Capital Flow from Rich to Poor Countries?*	解析小罗伯特·E.卢卡斯《为何资本不从富国流向穷国？》	社会学

An Analysis of Émile Durkheim's *On Suicide*	解析埃米尔·迪尔凯姆《自杀论》	社会学
An Analysis of Eric Hoffer's *The True Believer: Thoughts on the Nature of Mass Movements*	解析埃里克·霍弗《狂热分子：群众运动圣经》	社会学
An Analysis of Jared M. Diamond's *Collapse: How Societies Choose to Fail or Survive*	解析贾雷德·M.戴蒙德《大崩溃：社会如何选择兴亡》	社会学
An Analysis of Michel Foucault's *The History of Sexuality Vol. 1: The Will to Knowledge*	解析米歇尔·福柯《性史（第一卷）:求知意志》	社会学
An Analysis of Michel Foucault's *Discipline and Punish*	解析米歇尔·福柯《规训与惩罚》	社会学
An Analysis of Richard Dawkins's *The Selfish Gene*	解析理查德·道金斯《自私的基因》	社会学
An Analysis of Antonio Gramsci's *Prison Notebooks*	解析安东尼奥·葛兰西《狱中札记》	社会学
An Analysis of Augustine's *Confessions*	解析奥古斯丁《忏悔录》	神学
An Analysis of C. S. Lewis's *The Abolition of Man*	解析 C. S. 路易斯《人之废》	神学

图书在版编目（CIP）数据

解析巴鲁赫·斯宾诺莎《伦理学》: 汉、英 / 盖里·斯雷特（Gary Slater），
安德里亚·弗拉希米斯（Andreas Vrahimis）著；杨阳译. —上海：上海
外语教育出版社，2020
（世界思想宝库钥匙丛书）
ISBN 978-7-5446-6115-7

I.①解… Ⅱ.①盖… ②安… ③杨… Ⅲ.①斯宾诺莎（Spinoza, Benoit de
1632—1677）—伦理学—思想评论—汉、英 Ⅳ.①B563.1 ②B82

中国版本图书馆CIP数据核字（2020）第014383号

This Chinese-English bilingual edition of *An Analysis of Baruch Spinoza's* Ethics is published by
arrangement with Macat International Limited.
Licensed for sale throughout the world.
本书汉英双语版由Macat国际有限公司授权上海外语教育出版社有限公司出版。
供在全世界范围内发行、销售。

图字：09 – 2018 – 549

出版发行：上海外语教育出版社
　　　　　　（上海外国语大学内）　邮编：200083
电　　话：021-65425300（总机）
电子邮箱：bookinfo@sflep.com.cn
网　　址：http://www.sflep.com
责任编辑：梁瀚杰

印　　刷：上海叶大印务发展有限公司
开　　本：890×1240　1/32　印张 5.75　字数 194千字
版　　次：2020 年 9 月第 1 版　　2020 年 9 月第 1 次印刷
印　　数：2 100 册

书　　号：ISBN 978-7-5446-6115-7
定　　价：30.00 元
　　　　本版图书如有印装质量问题，可向本社调换
　　　　质量服务热线：4008-213-263　电子邮箱：editorial@sflep.com